► Lessons in Sustainable Development from China & Taiwan

DOI: 10.1057/9781137358509

Comparative Studies of Sustainable Development in Asia

Series Editor: **Sara Hsu**, Assistant Professor, SUNY New Paltz, USA

This series examines sustainable development in various countries in Asia, using a common framework with which to explore social, economic and environmental regulations and practices such as water pollution and consumption, income inequality and healthcare, and biodiversity. In each Pivot book of about 30,000 words, we explore the sustainable development frameworks of two countries, laying out their progress in this area and comparing the two to highlight policy recommendations.

The consistent sustainable framework applied to each country allows us to examine and compare sustainable development sub-topics across countries, and to clearly note the gaps in progress among countries. The common framework also allows us to highlight, among countries, which policies and experiments have been most successful.

Titles include:

Sara Hsu
LESSONS IN SUSTAINABLE DEVELOPMENT FROM CHINA & TAIWAN

DOI: 10.1057/9781137358509

palgrave▶pivot

Lessons in Sustainable Development from China & Taiwan

Sara Hsu

Assistant Professor, State University of New York, New Paltz

palgrave
macmillan

DOI: 10.1057/9781137358509

C.1

LESSONS IN SUSTAINABLE DEVELOPMENT FROM CHINA & TAIWAN
Copyright © Sara Hsu, 2013.

Corresponding author. hsus@newpaltz.edu. This monograph was supported by the Taiwan Fellowship awarded by the Taiwan Ministry of Foreign Affairs.

First published in 2013 by
PALGRAVE MACMILLAN®
in the United States—a division of St. Martin's Press LLC,
175 Fifth Avenue, New York, NY 10010.

Where this book is distributed in the UK, Europe and the rest of the world, this is by Palgrave Macmillan, a division of Macmillan Publishers Limited, registered in England, company number 785998, of Houndmills, Basingstoke, Hampshire RG21 6XS.

Palgrave Macmillan is the global academic imprint of the above companies and has companies and representatives throughout the world.

Palgrave® and Macmillan® are registered trademarks in the United States, the United Kingdom, Europe and other countries.

ISBN: 978–1–137–35851–6 EPUB
ISBN: 978–1–137–35850–9 PDF
ISBN: 978–1–137–32556–3 Hardback

Library of Congress Cataloging-in-Publication Data is available from the Library of Congress.

A catalogue record of the book is available from the British Library.

First edition: 2013

www.palgrave.com/pivot

DOI: 10.1057/9781137358509

To my mother, Joyce Keller

DOI: 10.1057/9781137358509

Contents

DOI: 10.1057/9781137358509

List of Illustrations

Figures

Tables

palgrave▶pivot

www.palgrave.com/pivot

Introduction

Abstract: *Both Taiwan and China are extremely populous nations in terms of population per square kilometer of land, and both, due to population pressures and continuing high growth levels, have experienced challenges to sustainable development. In this book, we examine China and Taiwan from a sustainable development perspective, in terms of inequality and environmental issues. On the whole, Taiwan's equality and its environmental statuses are much more favorable than that of China, as a result of its pattern of growth and its stage of growth respectively. China's equality situation is an outgrowth of its pattern of development and it is stark, particularly between the rural and urban regions, which has resulted from its urban, coastal-biased pattern of development. Implementation of social services in rural areas is essential, as is creating sustainable growth in rural regions. China's environmental situation is dire and requires accelerated attention.*

Hsu, Sara. *Lessons in Sustainable Development from China & Taiwan*. New York: Palgrave Macmillan, 2013. DOI: 10.1057/9781137358509.

In this book, we examine sustainable development in Taiwan and China. While we know a lot about the economic growth paths of Taiwan and China, scholars have written little about sustainable development comparing the two nations. We define sustainable development, in the context of this project, as economic development and growth that contribute to environmental sustainability (low or non-polluting) and to social sustainability (promoting economic equality, improving health, etc.).

Both Taiwan and China are extremely populous nations in terms of population per square kilometer of land, and both, due to population pressures and continuing high growth levels, have experienced increasing challenges to sustainable development. Some of these are more visible than others. A visit to Taiwan's Changbin wind farms might convince an outsider that greenhouse gas emission is not an issue. A stay in Shanghai's Bund area might lead a traveler to believe that China has little to no poverty, even though poverty remains a large issue in the country. In this project, we examine the challenges to sustainable development from a comparative perspective, asking the following questions:

▸ In what ways, and to what extent, is China's economic growth sustainable?
▸ In what ways, and to what extent, is Taiwan's economic growth sustainable?
▸ What can China learn from Taiwan in terms of sustainable development?
▸ What can Taiwan learn from China in terms of sustainable development?
▸ How could a partnership in improving sustainable development benefit both nations?

We first lay out a theoretical framework from which to examine various aspects of sustainable development in Taiwan and China, describe these components of sustainable development, then look at specific economic policies and programs that the countries have implemented, and finally draw conclusions about what is necessary for future study and policy-making in the area of sustainable development.

Sustainable development is critical to future growth because global warming and, in some areas, social exclusion are at the forefront of global challenges. As one aspect of environmental sustainability, scientists have shown that climate change is a real threat, and social scientists

DOI: 10.1057/9781137358509

studying climate change have shown ways in which environmentally neutral climate-change policies can reduce future costs due to global warming. Social exclusion, unlike climate change, has no upper bound at which reversibility is impossible, but increasing economic inequality has produced clear winners and losers of economic growth. Reducing the number of losers is essential to maintain peaceful and useful growth in many years to come. Inequality has been highlighted as a central issue particularly in China. Policies that put forth development that is conducive to social and environmental sustainability are essential to lay out, in the hopes that they may be adopted by governments in due course.

DOI: 10.1057/9781137358509

1
Theoretical Framework

Abstract: *The field of sustainable development is relatively new, but some of the ideas—that environment, individuals, and society should be protected—are not. These have been incorporated into several strains of theoretical approaches on the subject. In this chapter, we discuss these approaches and draw out our own theoretical approach to the subject.*

Hsu, Sara. *Lessons in Sustainable Development from China & Taiwan*. New York: Palgrave Macmillan, 2013.
DOI: 10.1057/9781137358509.

DOI: 10.1057/9781137358509

The field of sustainable development is relatively new, but some of the ideas—that environment, individuals, and society should be protected—are not. These have been incorporated into several strains of theoretical approaches on the subject. First, researchers have used systems theory to examine sustainable development. Systems theory, an interdisciplinary approach to analysis, examines sustainable development from a systems or structures perspective while analyzing elements within the system as they contribute to the structure as a whole (Bossel 1999). This perspective on sustainable development explores the relationships between variables and indicators and the way in which they affect viability of the larger environmental, social, and economic contexts.

Second, and relatedly, is the notion within complexity theory that is related to systems of requisite variety, which treats the system's ability to deal with future changes and adapt and respond to those changes (Nooteboom 2007; Ashby 1956). The theory of requisite variety states that policymakers must be able to respond to changes with as many different countering changes as needed to secure the system from the external environment. Systems must therefore be flexible and extremely adaptive.

Third, research has approached sustainable development from a transition management perspective, which states that sustainable development should be thought of as a dynamic process requiring management of transformation that prevents states becoming trapped in sub-optimal solutions (Kemp, Parto and Gibson 2005). The management process should attempt experiments in sustainable development, use transition agendas, evaluate the transition process, and maintain public support. The concept of a sustainability knowledge transition is related to the study of transitions management and finds that a country must move from acting as a "supply chain" to acting as a "knowledge chain" in order to become a more sustainable society.

Fourth, the Integrated Global Perspective finds that an integrated perspective relating social and natural environments to global change is necessary to bring about sustainable development (Choucri 1995). This theory also states that individuals and institutions alike need to take responsibility for what they can affect, in terms of production and consumption. Business and other transactions that are "business as usual" may still create environmental degradation.

Although these theories can be applied to our study of sustainable development in China and Taiwan, we look less at the process of how

DOI: 10.1057/9781137358509

components of sustainable development fit together and change in and of themselves, and more at how sustainable development can be thought of in terms of economic growth and development. We wish to point out that the notion of sustainable development has been used as a basis for changing the focus of economic development theory. Economic growth, which was used as the main emphasis in development theory before the nineties, was overshadowed by more comprehensive measures of economic development that included aspects of environmental condition, health, education and other social factors, and social and economic equality embodied in the Human Development Index and many other indicators that were created on this basis.

For this book, we wish to highlight the impact of economic growth on human well-being and environment as aspects of sustainable development. In some cases, there tends to be a trade-off between well-being and environment, as in China. Economic growth has led to pollution in cities, in which residents enjoy the highest levels of well-being. In other cases, there may be a correlation between poverty and environmental degradation, as the poor exploit their own resources in order to survive. The impact of growth on well-being and environment is less clear in Taiwan since this country has undertaken large efforts to improve both well-being and environment.

Geography is another factor that has influenced sustainable development. The field of new economic geography was pioneered by Krugman (1991). This field finds that geographic factors influence many aspects of economics. In this book, then, we find that there are two layers of causality. While geography influences economic structure, both factors influence effects of industrialization on the environment, and both also influence physical, social, and economic well-being (with inequality as one component of this). While the purpose of this book is not to lay out a theory and test it with empirical data, we can expand on this a bit to provide a framework for understanding the description we lay out in the rest of this text on sustainable development.

One is hard pressed to find papers that illustrate a theory quite like this. Geography, economic structure, well-being (including poverty and inequality), and environment are conceived of in the literature in the following stylized formulations: 1) poverty adversely impacts the environment, and adverse environmental effects exacerbate poverty and other aspects of well being, such as health (WCED 1987, Dasgupta 1995; Mabogunje 1995; Blaikie and Brookfield 1987; Mitchell and Popham

2008); 2) economic structure adversely or positively impacts the environment (McGregor, Swales and Turner 2004; Jänicke et al. 1989; Chen 2007); 3) economic structure adversely or positively impacts well being (including inequality) which varies geographically and may be impacted by geography (Bhalla, Yao and Zhang 2003; Chang 2002; Chen 1996; Jones, Li and Owen 2003; Kanbur and Zhang 2004; Sicular et al. 2005); and 4) geography influences economic structure (Krugman 1991; Fujita and Krugman 2003; Krugman 1999).

The diagram below illustrates how these factors influence one another.

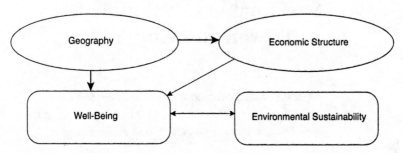

FIGURE 1.1   *The interaction of geography, economic structure, well being, and environment*

Source: The author.

Geography can influence well-being and economic structure. Economic structure may in turn affect environmental sustainability and well-being. Well-being and environmental sustainability affect one another. In what follows, we describe many facets of sustainable development, and we can bear in mind the ways in which these individual components of sustainable development are shaped by geography and economic structure, as well as the way in which they interact with one another.

These components are also shaped by humans—policies impact sustainable development and also are impacted by sustainable development. Hence we can talk about the natural and human forces that shape these components, but also about policy implications that arise from a study of these factors.

DOI: 10.1057/9781137358509

2
Sustainable Development in Taiwan and China

Abstract: *In this chapter, we discuss sustainable development in Taiwan and China in terms of geography, well-being, economic structure, and environmental sustainability.*

Hsu, Sara. *Lessons in Sustainable Development from China & Taiwan.* New York: Palgrave Macmillan, 2013. DOI: 10.1057/9781137358509.

DOI: 10.1057/9781137358509

Introduction to Taiwan

In Taiwan, income distribution is more equitable and the environment is in a better position than in China. However, over 80% of Taiwan's population lives in urban areas. Short-term returns in building up the cities has had an adverse impact on the environment (Huang, Wang and Budd 2009), which the country is attempting to reverse.

Economic development policy previously triumphed over environmental policy (Chen and Huang 1997). Increasing GDP was the main target, particularly in the 1960s, when Taiwan began a major focus on export-oriented industrialization. Manufacturing was dominated by the shoe industry in Central Taiwan in the sixties, the petrochemical industry in Southern Taiwan in the seventies, and the semiconductor industry in Northern Taiwan in the eighties (Hsu and Cheng 2002). None of these industries was particularly "green." As Taiwan climbed the technology production ladder, it gradually began to take into account its impact on the environment.

During its period of growth, however, Taiwan experienced growth in manufacturing, coupled with rapid urbanization. As a result, urban sprawl and landscape destruction have led to deforestation and pollution of air, water and soil (Tseng, Fang, Chen and Loh 2009). The relatively small geographical territory of Taiwan has been stripped of its natural resources as urbanization, increased vehicle usage, and factory density have increased. The population pressures on the land are becoming intense. Manufacturing shifted from being labor-intensive to being technology-intensive in the eighties, and most of the production activity remained in the cities. Urban areas remain locations in which air and water pollution are concentrated and must be effectively addressed.

There has been a constant conflict of interest between growth and protection in Taiwan, although the government has in the past 30 years begun to make legislation controlling waste and pollution. This is in large part due to increased public awareness and activism (Lyons 2005). Taiwan has adhered to the most basic international policies for protection of the environment, including UN- compliant transportation of radioactive waste, restrictions on trade in hazardous materials, and prohibition or regulation of persistent organic pollutants (Roam 2005).

After the Earth Summit in Rio de Janiero in 1992, Taiwan issued its own Agenda 21, National Report on Biodiversity, National Environmental Protection Plan, and a plan for developing Taiwan into a "Green Silicon

DOI: 10.1057/9781137358509

Island" (Council for Economic Planning and Development, Executive Yuan 2004). Of these, Agenda 21 is the most comprehensive voluntary international framework for sustainable development, focusing on social and economic dimensions, resource management and conservation, strengthening voices of major groups like women, children, and ethnic minorities, and how sustainable development can be effectively implemented. Taiwan's adaptation of these basic principles can be viewed below.

Taiwan's Agenda 21 basic principles

1   **Principles of Environmental Carrying Capacity and Balanced Consideration**
    Social and economic development should not exceed the carrying capacity of the environment. Environmental protection and economic development should be accorded balanced consideration.

2   **Principles of Cost Internalization and Prioritization of Prevention**
    Based on the principles that the polluter has the responsibility to solve pollution problems and the polluter/beneficiary pays, business sectors and society should internalize the external cost and reflect production cost rationally through market mechanisms and economic tools. Before the commencement of any activity that could possibly cause significant and unavoidable damage toward the environment, an environmental impact assessment should be implemented and efficient prevention measures adopted in advance, in order to minimize the damage.

3   **Principles of Social Fairness and Intergenerational Justice**
    Distribution of environmental, social and economic resources should accord with the principles of fairness and justice. This generation has the responsibility to maintain and secure sufficient resources to provide for the needs of future generations.

4   **Principle of Equally Stressing Technological Innovation and System Reform**
    Sustainable development strategies should be mapped out and the risks of policies evaluated, based on the spirit and method of science. Technological innovation should be considered as the driving force for strengthening both environmental protection and economic development. Decision-making mechanisms should be adjusted and related systems for implementing sustainable development should be established.

DOI: 10.1057/9781137358509

5   **Principles of International Participation and Public Participation**
To meet our responsibilities as a member of the international community, we should learn from the experience of the advanced industrial nations. Regarding the establishment of environmental regulations, we should follow international norms, give assistance to other developing nations, and set sustainable development as a priority objective. Decision making for sustainable development actions should be predicated on the views and expectations of people from all sections of society, with a transparent process of open and widespread consultation conducted as the means of forging consensus thereon. And the implementation of sustainable development policies should be carried out jointly by the government and private sectors, with each bearing its full share of responsibility and exerting its utmost effort to achieve the targeted results. (Source: Council for Economic Planning and Development, Executive Yuan (2004))

In order to better implement Agenda 21, Taiwan drew up its own program for sustainable development in 2002, which focuses on reversing some of the environmental destruction that has resulted from rapid growth. The program includes restoration of high quality forests and degraded habitat, restoration of wetlands, construction of water purification sites, establishment of a biodiversity database, and promoting voluntary greenhouse gas reductions (National Council for Sustainable Development 2010).

In addition, Taiwan's Environmental Protection Agency is an environmental institution in Taiwan that spends much of its budget on data collection. Laws regulate environmental protection in a variety of areas, including solid waste, water, and soil (EPA 2011). Taiwan also recently implemented a law taxing fuels and pollutants.

Taiwan and sustainable development

The EPA of Taiwan (2012a) uses the following framework for assessing its sustainable development progress, In terms of themes and sub-themes:

1   Environment (Air Quality, Water Quality, Waste, Environmental Management);
2   Energy Conservation and Carbon Reduction (Greenhouse Gas Emission, Energy Usage, Energy Conservation; Carbon Reduction);
3   National Land Resource (Land, Forest, Coasts, Water Resource, Natural Hazards);

DOI: 10.1057/9781137358509

4   Biodiversity (Heredity, Species, Terrestrial Ecosystem, Marine Ecosystem);
5   Production (Material Consumption, Cleaner Production, Agriculture, Fishery, Labor, Macroeconomic Effectiveness, Public Finance);
6   Livelihood (Water Usage, Transportation, Green Consumption);
7   Health (Medical Care, Nutrition, Health Risks);
8   Science and Technology (R&D, Telecommunications);
9   Urban and Rural Culture (Cultural Heritage, Community, Urban);
10   Well-being (Poverty, Income Equality, Social Welfare);
11   Governance (Crime, Education); and
12   Participation (International Participation, Public Participation).

We use the same framework, with some inclusions. We find this framework to be very inclusive of both human (individual and social) and environmental concerns. We add the following related aspects, including:

13   National resource accounting
14   Property rights
15   Energy self-sufficiency and international politics of energy markets
16   Implication for the rest of the world

As Yeh (2001) writes, sustainable development should be understood as creating a better institutional environment, rather than as a static outcome. In that sense, he also notes that it is important to recognize that Taiwan is an island nation with scarce natural resources, a colonial legacy, disaster-prone ecology, and trade dependence, which is unable to participate directly in most international organizations. Hence it is unique in its topography and political status, and sustainable development policies should attempt to improve institutions as well as address existing problems.

The country has performed very well in terms of sustainable development since the upgrading of the Environmental Protection Agency (EPA) to an agency rather than a bureau under the Ministry of Health (Yeh 2001). The EPA has improved many aspects of sustainable development and has taken an active role in promoting better environmental practices, to which we turn next. If we look at the categories individually in terms of progress, we can evaluate Taiwan's level of progress more clearly.

DOI: 10.1057/9781137358509

Environment

We begin with a discussion of the environment. Taiwan's geography has shaped the economic structure, with many small and medium sized enterprises producing goods and services within a limited geographical area. This has led to high levels of industrial agglomeration and high density living regions. In turn, controlling for environmental policies, this has had an adverse impact on air and water quality.

Air pollution

Good air quality days have increased while bad air days have decreased, particularly over the late nineties, as the EPA implemented air pollution fees, emission permits, and environmental impact assessments. Bad air days have higher levels of air pollution, which is a mixture of gaseous and particulate components (Hung et al. 2012). There is an extensive air pollution monitoring system in Taiwan, managed by the EPA. Implemented in 61 municipalities, the system captures daily readings of five pollutants, PM_{10}, O_3, CO, SO_2, NO_2. Taiwan is the 20th highest emitter of carbon dioxide (Chi 2010) and is currently in the process of reviewing a Greenhouse Gas Act. Greenhouse Gas reduction is right now voluntary.

There are many sources of air pollution. Motor vehicles emit fumes, including the chemicals benzene, polycyclic aromatic hydrocarbons (PAHs) and benzo[a]pyrene (B[a]P) (Hung et al. 2012). Air pollution traveling to Taiwan from China has increased as China's production has increased, with nitrogen oxide emissions in Mainland China increasing from 8.8 Tg year^{-1} in 1990 to 14.2 Tg year^{-1} in 2000 and SO_2 emissions increasing from 20Tg year^{-1} to about 27.5 Tg year^{-1} (Lin et al. 2005). A portion of these emissions travel by air currents to Taiwan. Air pollution also comes from domestic production, with factories emitting nitrogen oxides, sulfur oxides and volatile organic materials. PM10, or particulate pollution, is highest in Southern Taiwan in the winter and highest in Northern Taiwan between March and May due to dust storms in central Asia and the transport of dust by monsoons (Yang 2002).

The incentive system implemented beginning in 1995 to reduce air pollution has been successful in reducing pollution over the nineties and 2000s (Chang et al. 2010). When first implemented, pollution controls began with prohibition of higher levels of pollution, evolving to a comprehensive air quality management system. Air pollution fees require

DOI: 10.1057/9781137358509

polluters to pay for emission of sulfur oxides, nitrogen oxides, and particulate matter, and the fees are used by the EPA to pay for programs that improve air quality.

Emissions-control programs for stationary sources include a permit system, a fugitive emission control program, a clean fuel program and the enhancement of inspection program (Chang et al. 2010). The permit system for factories began in the 1990s under the Air Pollution Control Act, and specifies the amount of pollutants allowed for discharge. The fugitive emission control program controls production of organic solvents released by production processes such as dry cleaning and gas station operation. The clean fuel program controls SO_x emissions from fuel combustion processes. Finally, the inspection program strives to enhance monitoring of emissions. For mobile sources, emissions programs include the adoption of new exhaust emission standards for motorcycle, gasoline vehicle and diesel engines, a smoke check program and a high-pollution vehicle retirement program. The new emission standards and the smoke check program reduce the amount of emissions and smoke that vehicles can produce. Under the high pollution vehicle retirement program, motorcycles with two-stroke rather than four-stroke engines could be retired in exchange for cash back. All of these programs have aimed, from different angles, to reduce air pollution.

As a result of these programs, factories must be built using permits that specify pollutants that will be discharged (Chang et al. 2010). Over 95% of the factories had received permits by 2006. All stationary sources except for small businesses must periodically test and report emissions. Major power sources are required to continuously monitor emissions. Air pollution emission standards for 22 industries are enforced. The first vehicle emissions standards were put in place in 1987, while a smoke check program for diesel engines and motorcycles was implemented in the mid-nineties (Chang et al. 2010). Only unleaded gasoline has been sold since 1983.

Table 2.1 shows the emission of greenhouse gases, including carbon dioxide, methane, nitrous oxide, hydrofluorocarbons, perfluorocarbons, and sulphur hexafluoride, between 1996 and 2010.

From Table 2.1, we can see that levels of carbon dioxide emissions are high. Levels of methane and nitrous oxide emissions have declined over the period. Levels of hydrofluorocarbons, perfluorocarbons, and sulphur hexafluoride have fluctuated. Hydrofluorocarbons are used for refrigeration and air conditioning, while perfluorocarbons are used in

DOI: 10.1057/9781137358509

TABLE 2.1 *Emission of greenhouse gases by kind (10^3 M.T. CO_2 equivalent)*

	Total	Carbon dioxide	Methane	Nitrous Oxide	Hydrofluoro-carbons	Perfluoro-carbons	Sulphur hexafluoride
1996	208,193	175,729	15,495	14,217	2,752	NE	NE
1997	219,847	188,925	15,447	12,360	3,115	NE	NE
1998	229,760	198,312	15,149	11,908	4,391	NE	NE
1999	237,438	207,126	14,660	12,258	3,392	NE	NE
2000	256,612	224,621	11,028	12,444	5,639	2,386	494
2001	260,163	230,547	9,200	12,437	5,412	2,021	546
2002	267,547	239,575	7,250	12,205	5,415	2,509	593
2003	274,629	248,563	6,196	11,205	4,920	2,776	969
2004	283,470	257,185	5,920	11,734	4,494	2,852	1,285
2005	287,241	263,756	4,979	11,461	1,647	2,505	2,893
2006	294,526	271,688	4,486	11,674	1,028	2,657	2,993
2007	296,826	274,997	4,127	11,429	1,031	2,309	2,933
2008	284,498	263,589	4,727	10,839	1,001	1,498	2,844
2009	264,861	251,060	2,969	3,712	4,001	1,143	1,976
2010	277,618	265,078	2,194	3,071	3,941	2,318	1,016

Note: NE (Not Estimated) means that there is not enough data to gather statistics.

Source: National Statistics, Republic of China (Taiwan) (2013).

DOI: 10.1057/9781137358509

semiconductor manufacturing. Sulphur hexafluoride is used in the electrical industry for circuit breakers and other electrical equipment.

Noise pollution is another issue that affects many of Taiwan's inhabitants, who mainly live in large cities. A recent study on noise levels in Tainan finds that over 90% of inhabitants are exposed to unacceptable noise levels as defined by the US Department of Housing and Urban Development (Tsai, Lin and Chen 2008). These may include noises from vehicles, factories, and other businesses.

Water pollution

Sustainable water management is an issue in Taiwan, as floods and droughts increase and as water pollution remains a problem (Yang 2012). Tsai and Huang (2011) predict that future water resources will remain available in northern and eastern Taiwan but deteriorate significantly in southern and central Taiwan. This is due to patterns of water usage and reservoir capacity in these regions. Historically, reservoir and other water resource conditions have been better in northern and eastern Taiwan, and changes in water conditions in the future will only exacerbate differences between the north and east, and the south and central Taiwan.

Along with availability of water resources, water quality is another issue. The EPA monitors water quality across monitoring stations throughout the country. Monitoring is currently performed on 57 rivers, 60 reservoirs, 431 regional groundwater monitoring wells and 19 coastal areas. Taiwan's water quality survey in 2007 indicated that pH was of high quality standards while TP faced the lowest quality standards (EPA 2012c). All water samples across the country met standards for cadmium, arsenic, selenium and silver, while only 35% of them achieved the standard for manganese.

The major sources of water pollution include domestic sewerage, industrial waste water, and stock and farming waste water (Leu 2008). Construction of additional public sewer systems was undertaken to reduce river pollution. About 23% of domestic water drains into sewers, 14% goes into community sewer systems, 11% drains into built in sewage treatments in buildings, and other sewage goes into septic tanks. However currently, only about 48% of domestic wastewater is properly treated (EPA 2012b). The rest runs off into rivers, creating river pollution. In order to treat existing water resources, Taiwan Water Corporation was created in 1974, consolidating 128 water works across the country under

DOI: 10.1057/9781137358509

a corporate structure (Taiwan Water Corporation 2012). This is Taiwan's main water treatment and supply facility. Taipei, Taiwan exchanged its simple combined sewer (household sewer plus water runoff sewer) for separate sanitary and storm sewers to curb water pollution, which continues to be a serious problem (Chen and Lo 2010). This allowed for an increase in household wastewater treatment.

For industrial wastewater, the Water Pollution Control Act of 1974 focused on controlling wastewater treatment. More complete regulations were implemented in the nineties. Regulations on permissions, inspections, and licensing were thus put into place (EPA 2012b). The permission system included submission and approval of wastewater treatment plans before construction of factories and discharge of wastewater into the ocean and licensing of technicians. Inspections involve government supervision of factory operations and documentation of the results. For river water control, the government implemented the "Phase One of River and Ocean Management in Taiwan (2001–2004)," by which the government has removed pig farms and illegal metal processing factories and added an on-site treatment program particularly for the Erren River.

Everyday household and industrial pollution has not been the only source of concern. Environmental disasters have also shocked Taiwan's water resources. A large oil spill caused in 2001 by Greek merchant vessel Amorgos occurred after the Marine Pollution Control Act was enacted in 2000 and revealed some important lessons in responding to a marine disaster (Chiau 2005). The Marine Pollution Control Act set out ways to react to major marine pollution incidents. At the time, few institutions understood the Act and lacked professional manpower, facilities, and experience in ocean spill clean-up. The methods used during the clean-up also generated controversy. First, the response time was long and it was difficult to hire local people to cleanup the spill. This was exacerbated by weather conditions, including a monsoon. Second, oil that adhered to coral reef tidal trenches and pools also made the cleanup process more difficult, and it took time to address this issue. Third, because of adverse weather conditions, the ship was broken into several pieces, and part of it was left in the water at the command of the government and has continued to damage sea life. Fourth, cleanup efforts were not well integrated and it took time to build a command center to coordinate all of these efforts. These lessons were taken to heart by public officials, who quickly adopted procedures for dealing with this type of disaster in the future.

DOI: 10.1057/9781137358509

Other pollution

Heavy metal pollution has occurred in Taiwan's soil as well. Organisms in Kenting National Park were found to have heavy metal poisoning due to excess Cadmium, Mercury and Tin (Hsu, Selvaraj and Agoramoorthy 2006). These types of chemicals cause serious reproductive loss in birds. Heavy metals associated with urbanization and industrialization have found to be concentrated in the soils found in a number of regions. Heavy metals may linger for a number of years without dissipating on their own.

Energy conservation and carbon reduction

A recent study found that the only energy efficient cities in Taiwan are Kaohsiung City, Taipei City, and Chiayi City (Hu et al. 2011). The study uses Data Envelopment Analysis (DEA) in order to determine this. Regions that are less efficient use more inputs, such as electricity and oil, than they produce in terms of outputs, or real income. Part of the difference can be accounted for by heterogeneity in road density and educational attainment which can impact consumption of inputs, while some of the difference is sheerly due to better efficiency in particular cities. Changing the use of mainstream and alternative energy consumption are key components of improving the energy efficiency of cities.

Energy consumption

Energy conservation and carbon reduction are practices that reflect pro-environmental policies and movements. Like other nations, Taiwan struggles with energy conservation. Taiwan consumed 107.8 million kiloliters of oil equivalent (KLOE) of energy in 2004 (Chan et al. 2007), with almost all energy sources coming from fossil fuels. Since Taiwan is an island nation with limited natural resources, 98.7% of the energy required is imported (Kao 2008). Industry is powered in large part by coal and oil. Coal comprises 48.9% of energy resources, while oil comprises 32.3%, gas comprises 10.1%, nuclear comprises 8.3%, and renewables comprise 0.4% (Hu 2011). Neither coal nor oil represents clean energy sources.

Taiwan's industry, which accounts for about half of Taiwan's energy consumption, is made up of many small and medium sized enterprises that are increasingly adopting better energy management technology,

DOI: 10.1057/9781137358509

with government financial and technical assistance (Chan et al. 2007). Energy-intensive industries like iron and steel, chemical, cement, pulp and paper, textiles, and electrical use 90% of total industrial energy consumption. CO_2 is a major emission. These energy intensive industries could improve efficiency by using high efficiency equipment, recycling waste, and adopting new types of technologies.

Due to increasing emission controls, the average growth rate of CO_2 emission has slowed down since 2000. The Minimum Energy Performance Standard (MEPS) is the main regulatory tool used for energy efficiency, and requires manufacturers and importers to comply with energy requirements. Energy conservation labeling is a voluntary program to induce consumers to buy energy-efficient products. Because Taiwan has a less developed renewable energy sector, efficient energy usage and conservation is key to reducing greenhouse gas emissions. High energy-saving methods include using more efficient electrical motors, lighting facilities, refrigerators, air compressors, boilers and furnaces (Chan et al. 2007).

Alternative energy

In order to combat global warming, increase energy independence, and create an energy industry, Taiwan is attempting to develop new technology for alternative energy production. They possess an advantage in the manufacturing of photovoltaic cells for solar power, and in 2004 became 4th in the world in solar cell production. Though the landscape limits its ability to use wind power, Taiwan is producing an increasing amount of biodiesel and bioethanol. The development of these two energy sources depends on biomasses with a high sugar or starch content. Development of good alternative energy sources is essential to improving Taiwan's ecological footprint.

Taiwan is in the process of attempting to increase its renewable energy capabilities. Wind energy is the largest resource, followed by solar energy, biomass, ocean energy, geothermal energy, and hydropower (Chen et al. 2010). Conditions of wind power development are relatively good and can be enhanced with the development of wind power storage capacity. Taiwan's natural conditions are suited to solar energy, but land conditions are not. Taiwan receives a great deal of sunlight but is limited in terms of land mass on which to generate solar energy. Limited land mass also restricts the amounts of biomass energy that can be generated.

DOI: 10.1057/9781137358509

Several issues with renewable energy must be overcome before it can be more widely used: low energy density, high cost of power generation, instability of power supply, and current cost of renewable energy being.

National land resource

Although Taiwan's geographic area is small, its land resource is quite diverse in terms of natural landscape. Because it is an island nation in a tropical region, it is subject to typhoons and flooding. Environmental degradation has been exacerbated by economic activity on this sensitive land mass.

Taiwan is a country with abundant natural land resources that are increasingly threatened by climate change as well as by the effects of pollution and population. Taiwan is a sub-tropical island comprised of a central mountainous region with coastal regions on its perimeter. Although most of the land area is not arable, the nation hosts a wide variety of plant and animal life. The island is vulnerable to flooding and earthquakes which can affect both the highly dense population and the myriad plant and animal species (Council for Economic Planning and Development, Executive Yuan 2004).

Taiwan separates national land resources into urban planning land (12.5% of land) and non-urban planning land (87.5% of land) (Hsu and Lai 2007). The Taiwanese government has strived to reduce damage to its natural landscape by creating national parks and protected areas. Taiwan created its first national park in 1984 in Kenting (Lo et al. 2012). Currently, Taiwan has 6 national parks, 18 nature reserves and 24 nature protected areas, comprising 12.2% of its total land area. Taiwan is also attempting to control and reduce damage of its coastal environment.

Taiwan faces pressures on water resources due to growing demand for water. Climate change has also made the water supply more unstable, as global warming has resulted in more intense rain that causes reservoirs to accumulate silt (Gao 2009). Taiwan does have 151 large and small rivers, but most of these run from high mountains through short and steep courses and cannot be tapped as resources for drinking water (Liu 2011). Taiwan therefore has poor access to water resources despite above-average rainfall.

Major mineral commodities in Taiwan include marble, limestone, dolomite, natural gas, and petroleum (Tse 2010). Petroleum, natural gas and marble are the most valuable minerals. The continental shelf may contain large quantities of natural gas reserves (Britannica 2013a).

DOI: 10.1057/9781137358509

However, mineral resources are modest in quantity, and minerals such as recoverable coal, metallic minerals, and talc have been depleted. What is more, Taiwan lacks facilities for primary aluminum or refined copper production. Therefore, natural resources must be better protected and maintained in order to sustain the current level of living.

Biodiversity

Taiwan's wide range of climate zones has resulted in rich biodiversity, but industrialization has threatened many of the species. In order to combat this threat, national forests have been designated to ensure protection for wildlife and their habitats. In addition, the Taiwan Forestry Research Institute has set up long-term hydrology observation zones in forest watersheds and has studied aquatic chemistry, nutrient cycling and hydrological processes (Kuo 2011). The "Biodiversity Promotion Plan" put forward by the Executive Yuan in 2001 has sought to maintain biodiversity in Taiwan by protecting ecosystems, species, and habitats, and by requesting establishment of a national information center, an information exchanging mechanism, and databases for biodiversity. Currently, there are more than 10 institutes and museums that collect specimen data for Taiwan, and these are all integrated and searchable (Shao et al. 2007). Taiwan's endangered species are increasingly protected by the government, which imposes fines against abuse and bans fishing for endangered oceanic species (BBC 2011).

In addition, the Council of Agriculture created 19 nature reserves, 9 forest reserves, 14 wildlife refuges, and 28 wildlife habitats under the Cultural Heritage Conservation Law, the Taiwan Forest Management Plan, and the Wildlife Conservation Law (Council of Agriculture Forestry Bureau 2012). The Council of Agriculture subsidizes local governments to support wildlife conservation activities; 10 local governments have special departments to address conservation tasks.

Production

Production and the environment

Industrial production has adversely impacted the environment, and while policies have sought to counter pollution, a better solution for the

DOI: 10.1057/9781137358509

TABLE 2.2 Indices of forestry and fishery production (2006 = 100)

	Forestry Products			Fishery Products				
	Group Index	Saw Timber	Fire & charcoal wood and bamboo & byproducts	Group Index	Far-sea Fisheries	Offshore fisheries	Coastal fisheries	Aquaculture fisheries
1982	610.5	1889.2	197.7	83.2	44.4	280.8	80.6	81.3
1983	721.7	2409.3	162.2	85.5	44.9	264.1	88.5	91.0
1984	623.8	2056.9	150.5	90.7	51.3	268.0	96.6	92.8
1985	561.5	1846.3	137.7	94.6	56.4	264.1	104.0	95.5
1986	622.2	1986.8	181.8	100.3	63.5	241.9	111.0	105.0
1987	577.0	1626.2	232.1	113.9	74.5	252.6	106.4	123.1
1988	374.0	897.2	196.6	118.5	90.0	250.7	98.2	115.7
1989	264.2	531.1	169.1	112.6	94.6	257.0	97.5	93.3
1990	226.0	333.2	180.4	119.6	99.9	206.5	94.0	116.2
1991	211.7	217.4	210.5	110.7	92.1	195.7	77.7	109.1
1992	127.7	266.6	78.5	106.8	94.4	195.8	79.6	95.3
1993	133.4	191.6	113.0	113.4	114.7	176.8	73.2	95.1
1994	100.7	141.3	86.5	100.0	93.1	164.5	67.6	90.7
1995	117.5	199.0	88.8	104.2	98.0	176.0	78.7	90.4
1996	96.8	163.4	74.1	99.3	94.3	164.8	70.7	87.0

DOI: 10.1057/9781137358509

1997	108.9	169.8	89.5	101.3	105.9	157.8	70.8	79.6
1998	137.9	172.8	131.7	101.4	113.5	134.4	78.1	75.7
1999	115.9	98.3	131.1	99.3	113.3	128.6	71.7	72.9
2000	114.3	32.7	157.2	108.6	116.7	132.9	85.4	89.6
2001	94.7	43.4	122.3	110.4	117.5	132.0	91.4	96.3
2002	105.6	64.9	127.6	119.4	123.3	142.5	94.4	110.4
2003	115.0	38.5	156.3	127.8	133.9	135.3	115.4	118.8
2004	114.8	51.2	149.1	118.7	123.5	129.5	106.8	110.4
2005	89.8	71.4	99.7	116.3	120.3	132.6	98.1	108.3
2006	100.0	100.0	100.0	100.0	100.0	100.0	100.0	100.0
2007	68.5	42.5	86.8	111.1	111.1	87.0	97.1	106.2
2008	63.4	45.7	75.8	86.5	86.5	78.9	86.2	103.7
2009	64.4	40.1	81.5	81.1	81.1	82.0	82.5	86.7
2010	59.2	29.2	80.2	85.1	85.1	70.0	63.7	93.3
2011	60.9	37.9	77.0	84.9	84.9	82.8	53.3	87.3

Note: Data do not include Kinmen County and Lienchiang County before 2009.

Source: National Statistics, Republic of China (Taiwan) (2013).

DOI: 10.1057/9781137358509

environment and the citizenry is to engage in cleaner industrial production. Cleaner industrial production methods are known and encouraged. The problem is that cleaner industrial production in Taiwan is voluntary. However, the government has developed national-level award competitions for companies that are able to implement successful cleaner production strategies (Ashton et al. 2002).

Cleaner agricultural production is being developed, particularly in order to address problems created by land use of slopelands, application of herbicides and insecticides, presence of livestock waste, depletion and salinization of ground water, and creation of crop residual waste (Feng 1997). Taiwan's Agricultural Technology Foresight 2025 sought to perform research from 2008–2011 for mid-and long-term planning in agriculture, including planning for environmentally improved agricultural practices (Sun 2012).

Forestry and fishing are another component of "green" production. Taiwan's oceans are monitored regularly. A recent report based on such observation by the Taiwan Environmental Information Association showed that overfishing is a serious problem (United Press 2011). Currently, less than 6% of Taiwan's surrounding seas is protected by regulations. Taiwan is also the largest distance fishing operator in the Western and Central Pacific.

Table 2.2 includes the indices of forestry and fishery production, from 1982 and 2011, with 2006 as the index year.

Table 2.2 shows that production of forestry products have sharply declined between the eighties and nineties, due in large part to the decline in production of saw timber. While fishery product production has largely remained constant over the period, we see a decline in production of fishery products from offshore fisheries. The decline in offshore fishing harvests has occurred partly as a result of reduction in natural yield and state policies to reduce offshore fishing.

Taiwan also participates in eco-labeling programs, including the Green Mark Program initiated by the EPA in 1992, and the Energy Label and Water Conservation Labeling programs, operated by the Bureau of Energy and the Ministry of Economic Affairs and Water Resources Agency, respectively (Yu 2007). Taiwan also participates in the Green Construction Labeling program, begun in 2004 and operated by the Chinese Architecture Center (CAC) under the Architecture and Building Research Institute, Ministry of Interior Affairs, and the Energy Star Label begun in 2001 and operated by EDF under EPA/USEPA. The Taiwanese

DOI: 10.1057/9781137358509

government targets a certain percentage of procurement through green suppliers. In addition, Taiwan's annual Green Industry Trade Show provides a forum for green producers to display their products to potential buyers.

Production and labor

Human well-being also struggles under the current economic structure. Taiwan has laws to protect formation of labor unions, and has ratified ILO conventions dealing with discrimination (ICFTU 2006), but problems exist. Although women are formally protected by law, they are promoted less frequently and earn less than their male counterparts. Women are also not granted maternity leave, and are forced to quit their jobs due to marriage, age, or pregnancy. Many migrant workers continue to face a lack of labor rights and engage in forced labor. Migrant workers earn less than their counterparts as well, and are excluded from the workers' pension system.

Two legal regimes are currently in place to protect workers' rights. These include the Employment Service Act of 1992, which states that employers may not discriminate against workers on the basis of race, class, language, thought, religion, marital status, party affiliation, age, birthplace, one's provincial/county origin, gender sexual orientation, facial features, appearance, disabilities, and former membership in labor unions (Chiao 2008). The second legal regime includes the Gender Equality in Employment Act of 2002, which addresses gender discrimination issues in the workplace. These laws protect workers in multiple ways, but neglect workers that experience discrimination based on age and sexual orientation. Therefore there is room for future legal reform.

Public finance

Taiwan's taxes are obtained from business and personal income, customs duties, value and non-value added business taxes, commodity taxes, tobacco and alcohol taxes, and securities and futures transaction taxes (Chen 2008). An examination of trends in Taiwan's ratio of tax revenue to GNP shows that the number peaked in 1990 at 20% and declined to 12% in 2003. Tax revenue per person increased to NT $64,000 in 1998 and declined through 2004. Tax reduction has been a hindrance to government revenue and public finance, although tax reduction policies have been justified on the grounds that they stimulate business. Much of

DOI: 10.1057/9781137358509

Taiwan's public finance has been used either for economic development or for social welfare spending. Public finance until 2000 was dominated by economic development projects, while after 2000 and the change in political leadership, there was an increased focus on social welfare spending (Chan 2008).

Taiwan's fiscal policy has incorporated more government expenditure over time, and although tax revenues have increased, the need for public services has grown (Lee 2010). Several aspects of fiscal policy include expansion of domestic demand (through construction projects), improvements in tax efficiency and equity, and energy saving and reduction in carbon emissions.

Livelihood

Economic structure has impacted water consumption patterns, which in turn affect well-being. In addition, human activity and interaction with the economic structure have impacted transportation emissions, also in turn affecting well-being.

Taiwan's per capita water usage is higher than that in Europe and the United States, at 72 gallons per day (Kaye 2011). Much of the water is used in the agriculture and semiconductor sectors. The microelectronics industry itself requires large quantities of ultra-pure water (Lin et al. 2007). Although residents are able to purchase water at a very low price and therefore use more of it, some efforts are being made to use water more efficiently in the manufacturing sector. Firms in Hsinchu Science and Industrial Park are attempting to improve water efficiency.

Taiwan faces the threat of water shortages despite being surrounded by ocean water and experiencing frequent typhoons. The typhoons in fact often carry silt into the water supply and create a burden on the water purification system (Wang and Wang 2010). Lack of water due to insufficient rainfall also creates water shortages. Shortages are accompanied by an increasing demand for water as population and economic activities grow. Water scarcity is then one component of the water shortage problem, with the other part consisting of challenges in proper distribution, supply and management.

Transportation also reflects and impacts livelihoods. Transportation emissions produce high levels of emissions, including particulate matter (PM), ozone or volatile organic components (VOCs), carcinogenic

DOI: 10.1057/9781137358509

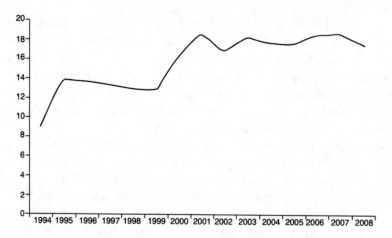

FIGURE 2.1   *Social welfare budget as ratio of central government budget (%)*
Source: Chan (2008).

toxics, carbon monoxide (CO), and sulphur dioxide (SO2) (Lan 2005). The number of privately owned vehicles has increased greatly since 1990, and these vehicles are used more than public transportation. Intercity public transportation passengers have declined since 1993 due to a large decline in bus passengers. Preference for private vehicles arose while availability of public transportation increased over the period. While sustainable transportation is available, individuals continue to prefer private, pollution-creating vehicles. Policy measures are necessary to change the public's preference for taking their own vehicles.

Health

Health as an aspect of human well-being has generally improved as Taiwan has grown wealthier. Policies to improve health care have played a large role in bolstering health. The National Health Insurance program, created in 1995, ensures access to health care for all residents (Kreng and Yang 2011). The National Health Insurance program is compulsory and insured individuals must pay premiums on time. The program is financed through premiums paid by individuals and taxes, and is a combination of public and private services (Cheng 2003). Over 90% of health care providers contract with the program. Individuals may choose

DOI: 10.1057/9781137358509

among providers without a referral. After the program was installed, the volume of services increased greatly, particularly in the areas of outpatient surgery, kidney dialysis, and emergency room visits. The program is generally characterized as easily accessible, with short waiting times and low cost, while quality of health care can be improved (Wu, Majeed and Kuo 2010). The health care system has allowed both poor and non-poor individuals to gain access to health care.

Some diseases that are not poverty related per se have faced different trends. Table 2.3 shows the number of patients of notifiable diseases, including tuberculosis, dysentery, typhoid and paratyphoid fever, acute viral hepatitis, severe enterovirus cases, AIDS, and dengue fever, from 1988 to 2011.

From Table 2.3, we notice several trends. First, several disease outbreaks spiked in the early 2000s, including tuberculosis, dengue fever, and dysentery. Second, the number of AIDS cases is on the rise. Third, typhoid and paratyphoid fever cases have remained constant over this period. Acute viral hepatitis and severe enterovirus cases have followed no particular trend. Taiwan's public health preparedness system has improved since the outbreak of SARS in 2003, including enhancements in the collection of disease surveillance data, preparation of a network of laboratories to make diagnostic tests, testing response systems to improve readiness levels, and investing in vaccine manufacturing capabilities in the case of an outbreak (Inglesby et al. 2012).

Science and technology

Many of Taiwan's science and technology policies have been geared toward enhancing and increasing economic production. Taiwan began its scientific and technological development in the sixties, importing foreign technology to aid production. The Industrial Technology Research Institute and Hsinchu Science Park aided the development of high technology, and by the eighties, technology research received the most focus (Science & Technology Policy Research and Information Center 2010). The Hsinchu Science Park offered foreign investors opportunities for research and development, providing a 5-year income tax holiday, a maximum income tax liability of 22% following the tax holiday, exemption from tariffs on machinery imports and from business tax on export sales, and venture capital assistance (Hung 2004). Taiwan's prowess in

DOI: 10.1057/9781137358509

TABLE 2.3 *Number of patients of notifiable diseases*

	Tuberculosis	Dysentery	Typhoid & paratyphoid fever	Acute viral hepatitis	Enteroviruses complicated severe case	AIDS	Dengue Fever
1988	6,501	192	98	161	NA	NA	1,938
1989	7,351	218	105	112	NA	NA	35
1990	7,329	141	81	90	NA	NA	10
1991	7,107	107	68	207	NA	19	175
1992	11,765	116	89	255	NA	22	23
1993	10,456	588	42	5	NA	34	13
1994	10,346	220	48	406	NA	66	244
1995	10,836	573	47	958	NA	100	364
1996	12,579	263	68	560	NA	162	53
1997	15,386	426	54	511	NA	135	76
1998	14,169	446	78	576	NA	155	334
1999	13,496	265	60	485	35	180	65
2000	13,910	643	46	298	291	181	140
2001	14,486	1,630	70	769	393	165	270
2002	16,758	725	72	806	162	177	5,388
2003	15,042	367	55	674	70	225	145
2004	16,784	252	57	807	50	257	427
2005	16,472	294	48	785	142	506	306
2006	15,378	264	53	613	11	579	1,074
2007	14,499	391	40	581	12	1,061	2,179
2008	14,424	317	44	631	373	850	714
2009	13,512	281	86	545	29	930	1,052
2010	13,393	434	45	344	16	1,087	1,896
2011	12,788	459	55	323	59	1,075	1,702

Source: National Statistics, Republic of China (Taiwan) (2013).

DOI: 10.1057/9781137358509

innovation was first centered around this science park, which focused on computers and their peripherals.

In the nineties, Taiwan continued to focus on improving technology, particularly in the Photonics, IT and communications, industrial automation, applied materials and biotechnology industries. In the 2000s, Taiwan focused on building technology in the semiconductor, IT hardware and software and broadband Internet industries (Science & Technology Policy Research and Information Center 2010). The National Science Council and Institute for Information Industry have launched technology development programs to boost Taiwan's competitiveness in innovation. Pilot programs to observe uses of new technology or to observe the environments in which new technologies can be developed have been promoted by National Science Council, Institute for Information Industry, National Taiwan University, National Chiao Tung University, and National Cheng Kung University, while the Ministry of Economic Affairs has promoted the Smart Living Technology & Service Program, aimed at developing Smart Towns and Intelligent Parks in Songshan, Nantou Puli, Yilan, Taichung, and Kaohsiung (Tsai and Lin 2012).

Taiwan's telecommunications industry is also competitive. Taiwan privatized its state owned telecommunications carriers in 1996 and opened up further to foreign investment (Li 2010). Taiwan's telecommunications sector has opened up to American competitors starting in 1998, as part of a bilateral WTO accession agreement (Lin 2003). Currently, over 100% of individuals in Taiwan have cell phones (some have more than one) (Shay and Partners 2011). and over 70% of households have access to the internet (Liang and Lu 2006).

Urban and rural culture

The Taiwanese minority aboriginal population contains 14 different tribes, while most of the population is Han Chinese (Yeh 2009). An emphasis on preserving culture came about relatively recently, in order to enhance social well-being. After Taiwan's lifting of martial law, in the nineties, the country underwent a period of "museumification" and "heritagization" (Chiang 2010). Historic preservation emerged as a means of post-colonial identity reconstruction. Cultural festivals were used to promote local crafts and heritage sites. Japanese occupation has been embraced as an

DOI: 10.1057/9781137358509

aspect of cultural heritage. Both urban and rural culture has preserved traditional cooking, festivals, and many types of religions, including Buddhism, Taoism, I-Kuan Tao, Catholicism, Christianity, and Islam.

Well-being

The impact of the economic structure on equality as an aspect of well-being has been large and positive. Taiwan has a very low amount of economic inequality. Although inequality rose in the eighties as industry shifted its structure to more skill-intensive processes (Lin and Orazem 2003), leaving out some of the unskilled workers, income inequality is lower than that of China, and its manufacturing development has unfolded more equitably than in many other nations. Its Gini coefficient was .34 in 2004 (Lee 2008), similar to that in Ireland, Spain, Greece and Poland (UNDP 2009).

In Taiwan, the urban–rural income divide is not present, in part because most residents live in urban areas, but also because Taiwan has institutions that ensure equity. Taiwan's Gender Equality Labor Law requires equal pay for equal work, maternity leave, and non-discriminatory hiring practices for married and pregnant women (Peng and Wong 2008). Senior citizens' welfare services are also in place. Although there are some issues regarding equitable compensation, Taiwan has done relatively well, and in this regard is comparable to many developed nations.

Li (2000) even makes the case that Taiwan is a model of equitable growth. The predominance of small and medium enterprises has allowed for more even distribution of income during industrialization. What is more, as Taiwan grew, it implemented a comprehensive Social Insurance Scheme, a universal health insurance program, and a pension program. Taiwan has increased its public education spending and provides free education for the first 9 years of schooling. Children of lower social status have access to schooling.

Taiwan also fares well in terms of fiscal policy and redistribution. In Taiwan, relative to developed nations, the share of taxes received from income tax is low but increasing, while the share of sales tax and property tax are declining (Ting-An Chen 2008). Individual tax burdens have declined since the 1990s and in 2003 were lower, as a percentage of GNP, than most countries in the world. Social expenditure in Taiwan has been 17 to 18% of the central government's budget in the 2000s, which is

DOI: 10.1057/9781137358509

similar to that of Japan, higher than that in the US, and lower than that in Germany and France (Chan 2008).

Governance

In the paradigm from which we are examining these facets of sustainable development, we consider government policies to be a "man-made" factor influencing well-being and environmental sustainability, as opposed to natural geography or market forces. However, policies interact with economic structure in particular. For example, crime rates have been impacted by government policy, and may also be driven by economic structure and social well-being. Education levels also interact with these factors.

Crime has increased since the lifting of martial law in 1987 (Mon 2001). Taipei County has the highest crime rate, with 183.48 crimes occurring among every 10,000 people in 1999. Following that were Taoyuan County, Hualian County, and Kaohsiung County. The overall crime rate is low. Table 2.4 contains information about violent crimes, including murder and non-negligent manslaughter, robbery and forceful taking, kidnapping, intimidation, and forcible rape, from 1982 to 2011.

From Table 2.4, we can see that cases of robberies and forcible rape have largely risen in the nineties and 2000s. Cases of murder, kidnapping, and intimidation have declined.

Taiwan's organized crime sector began to penetrate legitimate business in the early 1980s. After organized crime members killed a Chinese-American writer who had written a defamatory biography of the then-president, officials launched an assault on crime groups across Taiwan (Chin 2003). However, the lifting of martial law also increased organized crime. Some crime leaders became active in business and politics. Officials attempted to eradicate organized crime members in politics in the late nineties with limited success. In 1996, there were 1,208 crime groups with 10,346 members in Taiwan, of which 117 crime groups were considered a serious threat to law and order.

Education policies have also been laid out and to a large degree supported by the government. Taiwan's education system supports 22 years of education, with 2 years of preschool education, 6 years of elementary school, 3 years of junior high school, 3 years of senior high school or vocational education, 4–7 years of college or university, 1–4 years for a master's degree program and 2–7 years for a doctoral degree program

DOI: 10.1057/9781137358509

TABLE 2.4 *Violent crimes*

	Total Cases	Murder & nonnegligent manslaughter	Robbery and forceful taking	Kidnapping	Intimidation	Forcible Rape
1982	3,878	1,236	1,541	47	615	439
1983	4,385	1,441	1,745	91	653	455
1984	4,417	1,507	1,644	110	559	597
1985	5,542	1,362	2,501	69	843	767
1986	6,303	1,617	2,884	78	870	854
1987	6,095	1,572	2,569	77	992	885
1988	6,797	1,591	3,351	98	1,017	740
1989	7,835	1,699	4,429	160	942	605
1990	9,985	1,736	4,397	187	2,926	739
1991	8,683	1,784	3,385	128	2,620	766
1992	6,480	1,540	2,509	96	1,673	662
1993	7,110	1,622	3,092	106	1,418	872
1994	7,688	1,508	3,973	98	1,247	862
1995	16,489	1,765	11,670	98	1,817	1,139
1996	16,827	1,798	11,376	149	2,143	1,361
1997	13,648	1,712	8,169	92	2,197	1,478
1998	12,877	1,341	7,356	64	2,191	1,925
1999	11,362	1,269	6,328	63	2,019	1,683
2000	10,306	1,132	7,250	77	27	1,728
2001	14,327	1,072	10,959	76	35	2,125
2002	14,895	1,156	11,250	83	35	2,289
2003	12,966	1,057	9,236	73	45	2,445
2004	12,706	910	9,525	80	23	2,101
2005	14,301	903	11,022	65	31	2,235
2006	12,266	921	8,911	64	24	2,260
2007	9,534	881	6,059	38	24	2,487
2008	8,117	803	4,925	29	10	2,319
2009	6,764	832	3,799	18	5	2,073
2010	5,312	743	2,557	19	9	1,959
2011	4,190	686	1,661	10	5	1,800

Source: National Statistics, Republic of China (Taiwan) (2013).

(Ministry of Education 2011). Of this, 9 years are compulsory—6 years for elementary education, 3 years for junior high education. Higher education, particularly in engineering and the natural sciences, has played a positive role in Taiwan's economic development (Lin 2004).

Participation

International and public participation influence economic, social, environmental, and political policies. Increased public participation

DOI: 10.1057/9781137358509

strengthens policies that impact the economic structure, well-being, and environmental sustainability and promotes individual human rights. Broader domestic support for policies helps to ensure the well-being of citizens as they help to shape and comply with new regulations.

International participation

International participation is fraught with challenges for Taiwan due to its political situation. However, the nation has asserted itself as an international participant and strived to take part in higher level organizations. Currently, Taiwan is a member of 27 intergovernmental organizations, including the World Trade Organization, the Asia-Pacific Economic Cooperation forum, and the Asian Development Bank (Ministry of Foreign Affairs 2010). Due to its unique political status as a Chinese territory, Taiwan is not able to act as a member of the United Nations.

Public participation

Taiwan's public participation has fared better. Public participation is relatively high in Taiwan, with a democratic political regime and strong voter turnout. Taiwan's political regime, an electoral democracy, was rated "Free" by Freedom House in 2012 based on the existence of political rights and civil liberties (Freedom House 2013). Taiwan's high voter turnout shows election participation rates averaging 75% (Ho et al. 2013). Although corruption continues to be a problem, the media is free to report on the corruption.

National resource accounting

National resource accounting, or Green Accounting, is an indicator of environmental sustainability as it has been shaped by geography and economic structure. Green Accounting in Taiwan is compiled by the Directorate General of Budget, Accounting & Statistics (DGBAS) of the Executive Yuan in accordance with the basis cited in the UN's System of Integrated Environmental and Economic Accounting (SEEA) (Jao 2000). The basis for Taiwan's Green Accounting was created by the Environmental Protection Agency (EPA), which put together an Index for Sustainable Economic Welfare (ISEW).

DOI: 10.1057/9781137358509

Taiwan's Green Accounting accounts for natural resource depletion and environmental degradation in monetary terms under six major categories, including atmospheric environment, aquatic environment, terrestrial environment, ecological environment, human environment and miscellaneous environmental protection behaviors (Jao 2000). Natural resource depletion is further broken down into categories of depletion, namely, groundwater, gravel, natural gas, coal, and crude oil.

Despite the existence of Green Accounting, environmental degradation continues, with air pollution arising from automobiles and motorcycles, water pollution arising from industrial, residential, and agricultural discharges, and solid waste pollution arising from increases in population and economic activities. The Green Accounting matrix does not account for potential social costs, such as pollution costs to the human body.

Supplementing the Green Accounting data are reports from Taiwan's sustainable development department. The 2008 Annual Report on National Sustainable Development found that Taiwan's environmental quality is improving, but that carbon dioxide emissions are still not properly controlled (EPA of Taiwan 2008). The report also found that social pressure is increasing, with an increase in per capita garbage volumes, indicators of public health, including death due to cancer, have worsened, and public nuisance complaints have increased. Economic pressure indicators are moving in the direction of sustainability, with an increase in productivity of labor in manufacturing, and improvements in technology use. Institutional response has improved, with government response to environmental issues increasing, increase in positive planning and execution policies, and increases in government promotion of sustainability measures. Urban sustainability indicators are also improving, particularly in the area of urban per capita income, and also in terms of use of green space and public transportation. One adverse indicator in this area is an increase in urban car use. The Green Accounting indicators, together with the sustainable development reports, serve to provide a more complete picture of Taiwan's sustainable development status.

Property rights

Property rights policies have been shown to strongly impact both individual well-being and economic structure. Individual well-being is improved when individuals are allowed to own and trade in their own

DOI: 10.1057/9781137358509

property, while the economic structure is improved when individuals and institutions are allowed to protect and profit off of their own intellectual property. In Taiwan, individual property rights are protected, while the development of protection for intellectual property occurred over time.

During Taiwan's rapid growth process, the country faced pressure to improve its intellectual property rights regime, in particular by the United States. The Special 301 Provision of the 1974 U.S. Trade Act and other elements successfully influenced copyright enforcement in Taiwan (Chen and Maxwell 2007). Taiwan's Intellectual Property Office was established to enforce copyrights and educate the public about copyright infringement. Taiwan's prompt response to US concern for copyright protection has helped Taiwan to maintain its positive investment climate.

To enhance the innovation process, universities have been given some preference to patenting intellectual property over the government. Taiwan's Fundamental Science and Technology Act and Government Scientific and Technological Research and Development Results Ownership and Utilization Regulations that were passed in 1999 and 2000 respectively, give universities and non-profit institutions preference in ownership of their own inventions (Lo 2012). After these acts were passed, the number of patents granted to Taiwanese universities rose between 2004 and 2009. Many universities also set up intellectual property offices and internal policy to monitor intellectual property rights of work resulting from university-funded research.

Energy self-sufficiency and international politics of energy markets

Energy self-sufficiency is a high level policy that impacts economic and political institutions. Through these channels, it impacts environmental sustainability and social well-being as well. This is because the sources of energy produced and consumed can have variable impacts on a state, depending on the extent to which sources are emission-producing and economically/politically "sound." If sources are produced locally but highly polluting, for example, a policy of determined self-sufficiency may negatively impact the environment. In addition, if sources are produced abroad and, say, financed by a government deficit, they may also negatively impact government spending on social welfare. Hence

DOI: 10.1057/9781137358509

this high level policy can have complex effects on the dependent factors under study—namely, economic structure, environmental sustainability, and well-being.

Taiwan lacks its own energy resources but continues to desire energy self-sufficiency. The country is highly dependent on specific regions for energy, including the Middle East for crude oil, Australia and Indonesia for coal, and Malaysia, Qatar and Indonesia for natural gas. Taiwan's geographical isolation requires the island to retain higher generation reserve margins than in less isolated regions (Chang 2012). Taiwan's reserve margin is 28% and fossil fuels are used as back up reserves. In order to ensure a stable, ongoing supply of energy resources, Taiwan has set long-term contracts with suppliers, under the guideline that there must be an upper limit to the amount of energy procured from any one particular country or region (Chuang and Ma 2013).

Table 2.5 below contains various statistics on supply and demand of energy, including crude oil, natural gas, conventional hydropower, solar photovoltaic and wind power, solar thermal, coal, crude oil, liquefied natural gas (LNG), nuclear power, electricity, and heat, from 2007 to 2011.

Table 2.5 shows that trends in energy supply and demand have remained rather constant over the past 5 years, from 2007 to 2011. Taiwan is slowly trying to change this.

An unsuccessful attempt to increase nuclear power capacity occurred recently. In 2011, Taiwan's parliament approved a budget to continue construction on the fourth nuclear power plant in New Taipei City (Wakefield 2012). Public protests registered the public discontent with nuclear energy, particularly after a nuclear power plant melted down in Fukushima, Japan. The government, registering those protests, declared a policy to reduce nuclear dependence going forward.

Green energy may enhance energy self-sufficiency, but does not constitute a sufficient replacement for fossil fuels in this island nation. Green energy is targeted to represent 8% of electricity sources by 2025, but its creation is limited by Taiwan's small geographical territory and green energy's lower stability in comparison to fossil fuel energy creation. Nevertheless, policymakers continue to strive for some green energy creation and consumption through programs such as the creation of the "Office of Million Solar Rooftop PVs and Thousand Wind Turbines Promotion" (Ministry of Economic Affairs 2012) and the creation of low carbon cities, defined by the ratio of renewable energy consumption to

DOI: 10.1057/9781137358509

TABLE 2.5 *Energy supply and demand (10³ K.L. oil equivalent)*

	2007	2008	2009	2010	2011
Total supply	**146,095**	**141,169**	**138,048**	**144,550**	**140,675**
Indigenous Energy	959	912	875	893	947
Crude oil	18	16	16	14	11
Natural gas	371	318	312	263	293
Conventional hydropower	422	412	358	401	382
Solar photovoltaic & wind power	43	57	76	100	147
Solar thermal	105	110	113	114	113
Imported Energy	145,136	140,257	137,172	143,656	139,728
Coal and coal products	47,297	46,104	42,025	45,699	47,730
Crude oil & petroleum products	75,249	70,451	71,519	71,376	63,811
LNG	10,850	11,879	11,589	14,526	15,986
Nuclear power	11,740	11,823	12,039	12,056	12,201
Total demand	**146,095**	**141,169**	**138,048**	**144,550**	**140,675**
Domestic consumption	118,533	114,932	112,451	119,318	118,181
By sector:					
Energy sector own use	9,154	8,404	8,095	8,368	8,371
Energy use	105,335	103,222	100,849	107,263	106,272
Non-energy use	4,044	3,305	3,507	3,687	3,537
By final energy form:					
Coal and coal products	8,909	8,488	7,834	9,208	9,791
Petroleum products	48,553	45,994	46,774	48,405	44,921
Natural gas	2,419	2,493	2,497	2,967	3,413
Electricity	58,340	57,684	54,936	58,276	59,620
Solar thermal	105	110	113	114	113
Heat	207	163	296	347	322
Consumption per capita	**5,192**	**5,018**	**4,894**	**5,180**	**5,120**

Note: Data do not include Kinmen County and Lienchiang County except for 2011.

Source: National Statistics, Republic of China (Taiwan)

DOI: 10.1057/9781137358509

total energy consumption in rural areas and the greenhouse gas emission per capita in urban areas.

Taiwan is restricted in its own energy policy because of its tenuous relationship with China. Some recoverable oil reserves are believed to exist in the Tainan Basin of the Taiwan Strait, possibly amounting to 300 million barrels of oil and 41.7 trillion cubic feet of natural gas (Lai 2012). Some cooperation between Taiwan and Mainland China has occurred in the Taiwan Strait. However, in many cases there have been potential political conflicts between Taiwan and Mainland China over oil reserves, and in these cases, Taiwan has declined to be a party to discoveries of oil fields. Further, Taiwan is unable to participate in regional and international energy organizations due to its relationship with China. Due to its contested status as an independent nation, Taiwan remains unable to act as a recognized country participant in international energy organizations.

In order to improve Taiwan's energy security situation, Chuang and Ma (2013) recommend that China reflect these energy security costs through energy prices and improve energy security indicators by using renewable energy to hedge in favor of energy security. Taiwan would also benefit from a clarified energy policy with regard to China that provides some resolution toward the contested resources and an ability to participate in international energy organizations.

Implication for the rest of the world

Taiwan's government has increasingly strived to implement policies of sustainable development. While economically and socially the nation has performed in an increasingly sustainable way, the environment continues to lag. The difficulty that Taiwan faces in terms of changing its energy consumption to a greener composition reflects the problems inherent in many of the alternative types of energy sources and in the complexity of Taiwan's geography. This, coupled with Taiwan's high level of dependence on the outside world for energy resources, puts the nation in a precarious position energy-wise. Other nations that may be more energy self-sufficient face less pressure to change their energy composition.

Taiwan remains a model for income equality and citizens enjoy a relatively high quality of life. Compared to other nations, particularly to

DOI: 10.1057/9781137358509

China, Taiwan has developed in a way that allows individuals to gain access to equal opportunity of employment. The democratic nature of the political regime and the high levels of public participation help to ensure that human rights are protected. In terms of our conception of factors determining sustainable development, we can say that the economic structure has adversely influenced the environment, but very positively impacted well-being. Other nations can learn from Taiwan's egalitarian economic structure and implement this type of model, while controlling for adverse environmental impacts.

Introduction to China

Like Taiwan, China's main policy target for many years after reform began was economic growth. This is just now starting to change. Thirty years ago, China was a very poor country in which starvation was not unknown. At the outset of reform in 1979, most individuals were farmers, and those who lived in urban areas worked for state owned enterprises with little to no disposable income and no variety of goods to consume. After 1979, the farmers began to produce more as market systems opened up, but the real coup occurred when private firms began to spring up and become an increasingly common form of employment in the nineties. Here, China began its export oriented pattern of development, and the environment was the last concern as individuals faced the opportunity, often for the first time, to earn disposable income and attain a higher standard of living.

China's achievement of pulling over 200 million people out of poverty is incredible, and the nation could only accomplish this in such a short period of time by promoting growth over all other end goals. This was a celebrated victory for some time, as China became increasingly richer. However, as it became clear in the 2000s that growth was unbalanced, with some individuals gaining while others lost, and the victory lost its color. Citizens were dying from cancer in so-called "cancer villages" that were victims of industrial pollution. Forests were being eliminated at an alarming rate. Some individuals had no access to health care, no job security, and no real means of support. Other individuals became the nouveau riche and purchased luxury goods from abroad, while those in between moved into the middle class and bought cars, televisions, and cell phones. A wide variety of (un)sustainable development experiences arose.

DOI: 10.1057/9781137358509

China's sustainable development circumstances, like its economic, social and natural environment, are diverse. Some generalizations can be made. Chen and Lu (2008) capture a unique perspective: income inequality is tied to the spatial distance between inland and coastal regions, and to the concentration of economic growth in the latter. It also follows that industrialization in coastal regions leads to environmental degradation of those areas. If one considers the problem of sustainable development in China to be rooted in the pattern of economic growth, two questions must be asked: what policies have given rise to interregional disparities? And what patterns of growth and policies are in place that may have a countering effect on these disparities?

Chen and Lu (2008) list the policies that shaped interregional disparity in China (which is in turn at the heart of barriers to sustainable development):

▸ policies that allowed coastal regions to benefit from deregulation, allowing access to trade and marketization, and FDI (as found in Démurger and others 2002; Wan, Lu, and Chen 2007);

▸ growth of private economies and township-and-village enterprises in certain areas over others (Rozelle 1994; Wan 1998; Wan, Lu, and Chen 2007);

▸ fiscal transfers focusing on eastern areas (Ma and Yu 2003; Raiser 1998);

▸ variations in infrastructure in different regions (Démurger 2001); and

▸ industrial agglomeration mainly in or near coastal areas (Chen and others 2007; Lu and Chen 2006; Wen 2004).

These patterns have led to a vast amount of inequality, particularly between urban and rural areas. During the Maoist period, inequality between urban and rural areas was limited, as industrialization in urban areas was limited and health care and education in both urban and rural areas were provided. After reform and opening-up, all of this changed, with greater opportunity occurring in urban, coastal regions, and health and education retracted in rural areas. Income inequality grew between urban and rural areas, as well as between coastal and inland regions (Kanbur and Zhang 1999). Gaps between rural and urban areas grew after reform not only in income, but in health and education (Zhang and Kanbur 2005). Overall, the human development index (HDI) and principal components of the HDI currently reveal

DOI: 10.1057/9781137358509

that coastal regions are better off all around than inland areas, with Shanghai leading other regions, and Xizang (Tibet) trailing behind all other regions (Lai 2003).

Policies that may counter these disparities may include a focus on rural and inland development, changes in the fiscal structure, and improvements of infrastructure inland. China has attempted to reduce interregional disparity by improving infrastructure and spending inland. Currently, however, its focus, in reducing income inequality, is to urbanize rural residents, rather than to enhance rural development. Because this policy has only begun, we leave analysis of this policy for later work. Suffice it to say that this policy may only change the composition of income disparities. Below, we analyze in detail aspects of China's sustainable development.

China and sustainable development

China is still in the process of industrialization, and continues to be the factory of the world despite rising labor costs. It is in a stage that Taiwan passed through two to three decades prior, and therefore has not had much opportunity to develop environmentally sound practices. However, although Taiwan and China have different geographies, resources, and levels of development, we can apply Taiwan's sustainable development framework to China to draw a comparison and to determine exactly where China is in multiple areas of sustainable development. We start by looking at China's environment.

Environment

China's geography and economic structure have dramatically impacted the environment in an unsustainable manner. Because China's coastal region lies only in the eastern part of the country and permits large amounts of ocean freight travel, and because industry has developed around trade in this region, China's pattern of development has created massive amounts of pollution in its coastal area. Air and water pollution are both looming problems.

Air pollution

Air pollution is heavy in China. China's emissions account for one quarter of global carbon emissions, with per capita emissions above the

DOI: 10.1057/9781137358509

global average (Hallding, Han and Olsson 2009). Most of these emissions are due to burning of fossil fuels for energy. Further, one-third of emissions come from production of exports. China imports goods that are less carbon-intensive in production than it exports. Therefore, it is a net exporter of carbon. This is in part because the country is less carbon-efficient than developed economies.

Due to China's large population, per capita emissions are lower than those of Europe. At current emissions growth rates however, China will soon catch up to per capita emissions discharged by European residents. Since China represents a very large amount of global production, its control of emissions is critical for keeping global warming within the 2°C bracket (Hallding, Han and Olsson 2009). Emissions reductions are possible but will require drastic changes and further investment in low-carbon production.

Already, extreme air pollution problems exist in Beijing and Tianjin in central-eastern China, in Shanghai in the Pearl River delta, and in the mega-city of Guangzhou (Tie and Cao 2009). Air pollution is so severe that total suspended particulates in the majority of Chinese cities amount to twice the standard set by the World Health Organization, while sulfur dioxide emissions also remain exceedingly high (Diao, Zeng, Tam and Tam 2009). SO_2 and NO_X emissions in China cause acid rain, and nitrogen compounds cause eutrophication (over-fertilization) (Vennemo, Aunan, Lindhjem, and Seip 2009). The World Bank estimates that as much as 13% of all deaths in urban areas can be attributed to air pollution, which causes respiratory and cardiovascular diseases. CO_2 emissions from export production have increased as a percentage of all CO_2 emissions, underscoring the need for reduction of coal usage (Weber, Peters, Guan and Hubacek 2008).

Coal emissions are the dominant contributor to carbon emissions in China, at 80% of the total. Shealy and Dorian (2010) find that the only way for China to reduce carbon emissions, even with the addition of green energy resources, is to incur slower growth. In addition, aerosol emissions which include SO_4^{2-}, NO_3^- and carbon components due to anthropogenic activities and mineral dust pollution pose major threats to health (Tie and Cao 2009).

Table 2.6 provides information about ambient air quality in major cities in 2010. Air quality is quantified in terms of particulate matter, sulphur dioxide, nitrogen dioxide, and days of air quality greater than or equal to Grade II.

DOI: 10.1057/9781137358509

TABLE 2.6   *Ambient air quality in major cities (2010) (milligram/cu.m)*

City	Particulate Matter	Sulphur Dioxide	Nitrogen Dioxide	Days of Air Quality Equal to or Above Grade II (days)	Proportion of Days of Air Quality Equal to or Above Grade II (days)
Beijing	0.121	0.032	0.057	286	78.4
Tianjin	0.096	0.054	0.045	308	84.4
Shijiazhuang	0.098	0.054	0.041	319	87.4
Taiyuan	0.089	0.068	0.02	304	83.3
Hohhot	0.068	0.046	0.034	349	95.6
Shenyang	0.101	0.058	0.035	329	90.1
Changchun	0.089	0.03	0.044	341	93.4
Harbin	0.101	0.045	0.048	313	85.8
Shanghai	0.079	0.029	0.05	336	92.1
Nanjing	0.114	0.036	0.046	302	82.7
Hangzhou	0.098	0.034	0.056	314	86
Hefei	0.115	0.02	0.03	310	84.9
Fuzhou	0.073	0.009	0.032	351	96.2
Nanchang	0.087	0.055	0.042	343	94
Jinan	0.117	0.045	0.027	308	84.4
Zhengzhou	0.111	0.053	0.046	318	87.1
Wuhan	0.108	0.041	0.057	284	77.8
Changsha	0.083	0.04	0.046	338	92.6
Guangzhou	0.069	0.033	0.053	357	97.8
Nanning	0.069	0.028	0.03	349	95.6
Haikou	0.04	0.007	0.015	365	100
Chongqing	0.102	0.048	0.039	311	85.2
Chengdu	0.104	0.031	0.051	316	86.6
Guiyang	0.075	0.057	0.027	343	94
Kunming	0.072	0.04	0.046	365	100
Lhasa	0.048	0.007	0.021	361	98.9
Xi'an	0.126	0.043	0.045	304	83.3
Lanzhou	0.155	0.057	0.048	223	61.1
Xining	0.124	0.039	0.026	312	85.5
Yinchuan	0.093	0.039	0.026	332	91
Urumqi	0.133	0.089	0.067	266	72.9

Source: National Bureau of Statistics (2012).

From Table 2.6, we can see that the cities with the worst sulphur dioxide pollution are Urumqi, Taiyuan, Shenyang, Guiyang, and Lanzhou. Cities with the worst particulate matter include Lanzhou, Urumqi, Xi'an, Xining, and Beijing. Those with the highest levels of nitrogen dioxide pollution include Urumqi, Beijing, Wuhan, Hangzhou, and Guangzhou. Finally, those with the lowest number and percentage of breathable air days include Lanzhou, Urumqi, Wuhan, Beijing, and Nanjing. From

DOI: 10.1057/9781137358509

this information the overall worst cities in terms of air pollution include Urumqi, Lanzhou, and Beijing.

Water pollution

China also struggles with water pollution. Sixty percent of all rivers in China are Class IV or worse according to China's surface water quality standard, which means that humans must avoid direct contact with the water in these areas (Vennemo, Aunan, Lindhjem, and Seip 2009). Polluted water is used to irrigate crops in many instances—half of the rice yield, for example, is polluted with mercury, cadmium and lead. Groundwater is also becoming rapidly depleted.

The Huai River has experienced major pollution, and the central government has closed down thousands of factories that contribute pollutants to the river (Wang, Webber, Finlayson, and Barnett 2008). Yet water pollution remains a looming problem. China's rural development is unsustainable particularly in its consumption and pollution of water (Wang, Webber, Finlayson, and Barnett 2008). Non-agricultural rural enterprises can generally be characterized by their outdated technology and equipment, poor management, and large consumption of water and other resources. Most rural industries do not have wastewater treatment, and the worst polluters are papermaking, cement, and brick plants.

In rural areas, town and village governments often back pollute rural enterprises since the governments may be guarantors of enterprise loans and do not wish them to fail. Rural industries contribute the lion's share of local revenues. Local officials are not held accountable for pollution control, and local environmental protection bureaus are subordinate to the dictates of local officials.

Coastal regions are also strongly affected by pollution. Pollutants can enter the sea through river input, atmospheric deposition, industrial waste disposal, and terrestrial runoff. The concentration of heavy metals in coastal sediments and marine bivalves has increased over the past 10 years, and is of significant concern in terms of seafood consumption (Pan and Wang 2012). Metals are non-degradable pollutants and can heavily accumulate in coastal sediment. Bohai Bay in northern China is one of the most contaminated coastal regions in China. Other contaminated coastal regions include Jiaozhou Bay in northern China, the Yangtze River Estuary in eastern China, Hong Kong in southeastern China, and the Pearl River Delta in southeastern China.

DOI: 10.1057/9781137358509

Soil pollution is also problematic. Concentration of heavy metals in urban soils presents a health risk. Heavy metals can be inhaled, ingested, or absorbed by the body. Traffic, industrial, and domestic emissions all contribute to concentration of metals such as chromium, nickel, copper, lead, zinc and cadmium (Wei and Yang 2009). In addition, nitrogen pollution of soil, water and air is a growing problem. Nitrogen pollution of soil due to fertilizers and excreta and nitrogen emissions in the air are cause for great concern (Liu et al. 2011). Excess nitrogen affects soil acidification, plant growth, litter decomposition, greenhouse gas fluctuation, biodiversity, and ecosystem carbon exchange.

China has suffered from a number of pollution accidents. Between 2002 and 2006, provinces that experienced the highest number of accidents were Hunan, Jiangsu, and Zhejiang provinces (Hou and Zhang 2009). Half of the 82 recorded pollution accidents in this period were air or water pollution events. Half were associated with production accidents mainly resulting from technical, rather than human failure. Health damage, including death in some cases, resulted from half of these incidents.

China's 1983 law entitled Agenda 21 set environmental protection into law, focusing for some time on pollution control and currently both on pollution control and to some extent on ecological conservation (Zhang and Wen 2008). China also has a loose body of laws for environmental sustainable development, including the Law of Environment Protection, the Law of Water, and the Law of Saving Energy (Fan 2009). Government regulations aim at directly governing energy use and efficiency. However, quality standards and monitoring are low. In 2009 all provinces completed climate change plans, although no targets or decisions mandating climate impact reduction have been passed so far (Gemmer, Wilkes, and Vaucel 2011).

Energy conservation and carbon reduction

China has recognized that its pattern of development has been detrimental to the environment, and therefore has implemented policies to combat excessive energy use. China has been setting forth environmental Five-Year Plans since 1996, and is in the process of expanding its renewable energy resources. To reduce energy demand, the Chinese government required a 20% energy reduction between 2006 and 2010, which was to be carried out by the National Reform and Development

DOI: 10.1057/9781137358509

Commission (Zhou, Levine and Price 2010). Various policies for energy conservation were thereby enacted during this period. The policies included tax reductions for energy efficient technologies, national energy efficient design standards for buildings, appliance standards, and transportation standards such as emissions caps. Currently, 46 energy efficiency standards are in place for home appliances and commercial equipment (China UN Mission 2012).

However, China has missed its target of doubling coal use while quadrupling GDP between 2000 and 2020, surpassing the double-coal use target in 2006 (Shealy and Dorian 2010). The implications were that efficiency gains from reduction or improvement of state owned enterprises had been made before this period. Electricity consumption, including residential energy consumption, also increased due to rapid urbanization. Per capita residential electricity consumption for urban households increased over 7 times between 1985 and 2008, while tripling for urban households between 1979 and 2007 (Hubacek, Feng and Chen 2012).

China's carbon emissions have continued to increase, while carbon emissions per unit of GDP declined in 2011 (Johnson 2011). Carbon emissions are high due to China's continuing reliance on coal for power. In addition, although cleaner-burning coal plants are available, many Chinese firms are unwilling to pay additional costs to purchase them (Busby 2010).

Increases in construction of residential and commercial buildings have led to additional energy consumption. Heating and air conditioning account for most of building energy consumption (Yao, Li and Steemers 2005). Energy standards for heating new residential buildings have been developed by the China Academy of Building Research and were approved as industrial standard JGJ26–95 in 1996, while energy standards for heating existing standards were approved as industrial standard JGJ 129–2000 in 2001. Energy consumption for space heating is controlled at particular levels accounting for architectural and heating design under the Energy Conservation Design Standard. Industry standards for heating and cooling were put forth in 2001 under JGJ 134–2001 published by the China Academy of Building Research and Chongqing University and issued by the Construction Ministry. What is more, the Energy Conservation Law of 2008 applies to industrial facilities, commercial and residential buildings, the transportation sector, and public institutions (Shen et al. 2012). Going forward, energy conservation and carbon reduction are imperative, and additional

DOI: 10.1057/9781137358509

regulation and improved enforcement of existing regulation can assist this goal.

National land resource

China's natural geography is diverse. The way in which the economy and humans interact with the natural environment, however, has threatened the viability of some land and water resources. This adversely impacts human well-being. In this section, we discuss land use, land resource, water resource, and natural hazards.

Land use

Land use in China has created ecological problems, including erosion, desertification, and land pollution (Li and Zeng 2010). Land ecological security indexes have been created to gauge the extent of these problems based on the Pressure-State-Response Model recommended by the World Bank. This model states that humans exert pressure on the environment, changing the state of the environment. Society respond with policies that strive to reduce pressure on and improve the state of the environment.

China has not responded well to environmental pressures, although there is a diversity of land uses in different locations. Land use change has been either concentrated, as in Shanghai, or diverse (sprawling), as in Nanjing (Zhang, Uwasu, Hara and Yabar 2011). Looking at sustainability scores on 16 cities in the Yangtze River Delta, between 2000 and 2005, Zhang et al. find that cities that experienced concentrated but moderate growth had a better effect on the environment than did cities that experienced concentrated but rapid growth and cities that experienced diverse growth across a large area of land. Cities with concentrated but moderate growth conserved arable land, and had larger manufacturing sectors and better infrastructure. This may indicate that there is a preferred type of land use and economic development that responds better to environmental pressures.

Land resource

Despite possessing a large geographical area, China is below the world average in terms of forest coverage (Démurger, Hou, and Yang 2009).

DOI: 10.1057/9781137358509

China's forested area per capita is 0.13 ha, while the world average is 0.65. China is attempting to move away from a strategy of timber production to one of conservation, particularly with the introduction in the late nineties of the Six National Key Forest Programs. These include the Natural Forest Protection Program, the Sloping Land Conversion Program, the Desertification and Dust Storms Control Program in Beijing and Tianjin Municipalities, the Forest Shelterbelt Development Program in environmentally fragile regions, the Wildlife Conservation and Nature Reserves Development Program, and the Fast-growing and High-yield Timber Plantations Program. The most important of these are considered to be the Natural Forest Protection Program, which has focused on reducing annual timber production in natural forests, and the Sloping Land Conversion Program, which aimed to reduce soil erosion by converting cultivated land (including highly sloped land) into forests and grasslands. Together, these two programs have helped to improve China's forest resources.

China has also made use of its mineral resources. The nation is a large exporter and importer of minerals. China produces large amounts of aluminum, antimony, barite, bismuth, cement, coal, fluorspar, gold, graphite, iron and steel, lead, magnesium, mercury, molybdenum, phosphate rock, rare earths, salt, talc, tin, tungsten, and zinc (Tse 2010). China imports chromium, cobalt, copper, iron ore, manganese, nickel, petroleum, platinum-group metals, and potash. Extraction of these mineral resources has been relatively environmentally unsustainable, and reorientation toward a sustainable approach is being established. The mining industry in particular was notoriously unsafe, flouting safety and environmental regulations. In 2010, the National Development Reform Commission ordered the non-ferrous metals, iron and steel producers to shut down temporarily in order to reach energy goals of the Eleventh Five Year Plan. The Twelfth Five Year Plan directly targets sustainable development.

Water resource

China has a long coastline of 18,000 kilometers, which plays a very important economic role (Lau 2003). However, administrative division of the coastal region into 12 units creates an obstacle to establishing an integrated coastal zone management. Problems exist in terms of what the United Nations recommends for integrated coastal zone management, including problems with coordinated legislation, efficient institutional

DOI: 10.1057/9781137358509

organization, and a high degree of public participation. Red and green tides, caused by pollution and overfishing, have been caused by mismanagement of coastal zones, damaging water resources.

Water resources are threatened due to water pollution and increasing scarcity, resulting from demands of the large population and growth, as well as from poor implementation of environmental protection laws and poor water management (Turner 2007). In 2006, a survey conducted by China's State Environmental Protection Administration revealed that nearly half of the 21,000 chemical plants along the Yangtze and Yellow rivers are close to drinking water supplies. China's 14,000 factory farms also contribute greatly to pollution. Much of the municipal wastewater is not treated at all, and the situation is even worse in rural areas.

China's yearly per capita water supply is already 25% lower than the world's average (Turner 2007). This is especially a problem in northern China, which is a dry region. Water has been redirected from the South to the North through the South-North Water Transfer Project. China has also spent $3 billion in recent years to bring safe drinking water to 71 million rural residents. Groundwater is quickly being depleted as overpumping and contamination occur (Gleick 2010). As groundwater is lost, cities must dig even further to find clean drinking water. Given the number of threats to China's water resources, improved water resource policies may have a strong positive impact on China's water supply.

Natural hazards

China has a large amount of hill and plateau region, much of which is desert. China is very prone to natural hazards, such as floods, droughts, earthquakes, mudslides, typhoons and monsoons (Shi et al. 2007). Since the country lies in a major seismic belt, it experiences frequent earthquakes. Droughts occur in the Northwest Loess Plateau and the North China Plateau in the spring and autumn, while floods occur in the 7 large river basins, especially in the middle and lower reaches of Yangtze River and Huaihe River in the summer and autumn. Desertification approaches one-third of China's land territory. About one-third of China's total farmland area is affected by natural disasters annually.

Industrial agglomeration in coastal areas has resulted in environmental degradation in these regions. The ecological footprint, or the human demand on ecosystems, in Beijing, for example, was 4.99 global hectares

DOI: 10.1057/9781137358509

per capita, while the footprint in China as a whole was 1.78 global hectares per capita (Hubacek, Guan, and Wiedmann 2009). This is because in Beijing, demand for resources for both production and consumption purposes is high. China's megacities suffer from environmental degradation and likely impact the creation of natural hazards by altering global climate patterns through the process of global warming. Policymakers are attempting to address China's role in global warming, but with a very full economic agenda, this has become only one in a list of top priorities.

Biodiversity

China not only contains a diverse geography, but also contains a rich biodiversity that is threatened by economic and human activity. China is home to more than 30,000 plant and 6,300 vertebrate species (Tang et al. 2006). Many species have disappeared from lowland areas due to human activity but remain in the mountainous regions. Based on the criteria developed by the International Union for Conservation of Nature, over a third of animal species are endangered, while about three-quarters of plant species are threatened (Klok and Zhang 2008).

China's marine biodiversity consists of 20,300 recorded plant and animal species, living in four large marine ecosystems, including the Yellow Sea, the East China Sea, the Kuroshio Current, and the South China Sea (Qiu et al. 2009). China currently has 158 marine protected areas, covering a total of 3.77 million hectares of coastal and marine territory. However, the marine protected areas are not sufficiently protected. Large animals living in the Yangtze River have been threatened with extinction, including the Chinese paddlefish, the finless porpoise, and the Yangtze River dolphin (Lim et al. 2008). Marine biodiversity is threatened by the building of large dams, over-collection of species by local residents, pollution, and construction of harbors.

China has joined many international conventions and agreements with regard to biodiversity, including UNESCO's Man and Biosphere Program, the World Heritage Convention, the Convention on International Trade in Endangered Species of Wild Fauna and Flora, and the Migratory Bird Conventions (Klok and Zhang 2008). China has also prepared and coordinated a National Biodiversity Conservation Action Plan, which was ratified in 1994. Additional conservation action plans have since been created,

DOI: 10.1057/9781137358509

including the National Ecological Conservation and the China Hydrobiology Conservation Action Plans. Protecting biodiversity on the ground has not been as easy as implementing plans, however. For example, local governments manage most national reserves but are allocated minimal financial resources by the central government with which to do so. For this reason, many reserves lack management teams and staff. Therefore protection of natural species must be accompanied by funds to back the protection, as well as conservation education on how to do so (Liu et al. 2003).

Production

China's economic structure has neither been favorable to the environment, as industrial and agricultural production remain relatively "dirty," nor to human well-being directly, as labor relations and fiscal redistribution remain inequitable. The production regime has thus generally been far from sustainable.

Cleaner production

Cleaner production is one area of sustainable development that can be greatly improved. Some barriers to environmentally sustainable development exist at the firm level. For example, SO_2 emissions could be eliminated with the addition of an end-of-pipe flue gas desulfurization intervention that costs a power plant $40–$65 per kW. However, relative to polluter-pays costs for SO_2 emissions, the intervention is too expensive (Vennemo, Aunan, Lindhjem, and Seip 2009). Although this would have enormous positive externalities, this intervention has not been implemented by most firms.

In fact, some businesses, particularly Chinese equity joint ventures with partners in Hong Kong, Macao and Taiwan, are attracted to doing business in China because of weak environmental standards (Dean, Lovely and Wang 2009). Producing in a "dirtier" manner is far cheaper. This has resulted in 80% of foreign direct investment being located in eastern coastal provinces.

Agriculture

Agricultural production in China faces challenges to sustainable development. Fertilizer is often overused, creating water pollution, water is used

DOI: 10.1057/9781137358509

in irrigation and faces depletion, and methane emission contributes to global warming (Wang, Huang and Rozelle 2010). Agriculture is likely to impact and be impacted by climate change. With less water for irrigation due to water depletion and changes in rainfall patterns (causing floods or droughts), agricultural output will be strongly affected.

Material consumption

Production is also related to consumption. China's consumption has increased, as incomes have also risen mainly in urban areas. The young urban population is increasingly eager to consume (Stein 2009). Automobiles are more common and consumer durables such as refrigerators, air conditioners, computers and cell phones are more widely used. Meat consumption has also increased, resulting in deforestation and increased water and energy usage for the breeding of livestock. Consumption is viewed as an important activity for Chinese citizens, while consumer democracy is viewed as an opportunity for citizens to express themselves.

Labor

Labor is an important component of production. Mass protests over labor rights have increased since reform and opening up, as have the number of cases brought before labor arbitration committees (Bartley and Zhang 2012). A major labor law passed in 2007 had the effect of making workers more aware of their own labor rights, and the even more general government policy of turning away from export led growth and toward domestic led growth has placed a greater emphasis on higher wages, so that workers can afford to purchase the goods they produce. Although the Labor Contract Law Labor of 2007 did not increase wages, it increased the percentage of workers with contracts and reduced violations of workers' rights (Li 2011). Labor standards certification initiatives have sought to enhance corporate responsibility toward workers.

However, violations of labor rights remain a problem in China. Reports of worker suicides at Foxconn, an Apple subcontractor, in 2010 , revealed violations of labor rights, including excessive overtime, dangerous working conditions, and unsanitary food in cafeterias. These events drew international attention to Chinese labor conditions. Sweatshop conditions throughout China have persisted since the export boom began, and those who are least protected by law, particularly rural migrants, have faced often squalid working environments.

DOI: 10.1057/9781137358509

China's export sector has demanded young, rural migrant labor, which it can pay extremely low wages. While there is currently excess supply— i.e. unemployment—in the group of rural migrant labor over the age of 35, there is excess demand on the part of employers for young migrant workers (Chan 2010). Young workers, however, coming from one child families, are more likely than older workers to seek jobs in which they find dignity rather than just a paycheck (Wang 2012). Hence the labor market for young workers has changed, as workers assert their right to decent employment.

Macroeconomic effectiveness and public finance

China's macroeconomic policy has had varying degrees of effectiveness in controlling the production environment. China's government has used monetary policy in order to control inflation and maintain price stability. Monetary policy is not used to promote economic growth and been weakly effective in controlling price stability; often the relationship between the money growth target and the actual have been disparate (Geiger 2008), due to inaccurate timing and excessive money supply outside of the banking system.

Fiscal policy has been somewhat more effective. China has implemented fiscal policy where necessary to support major policy initiatives, and more recently, to combat the downturn caused by the global financial crisis. Fiscal policy has faced a big shift since reform, shifting from support of state owned enterprises to larger infrastructure projects. It has served to reorient the economy toward growth policy. Fiscal policy used during the last downturn (the global financial crisis) was largely effective, combating the worse effects of the crisis. Without it, China's economy would likely have substantially slowed due to a decline in export demand from abroad.

China's tax-sharing reform in 1994 increased the share of government spending as a percentage of GDP and increased the share of taxes going to the central government (Wang 2008). Over time, China has developed a better tax base from which to obtain fiscal funds. At the same time, China has improved and increased spending on public education and decreased spending on health and healthcare.

However, some local governments receive more than others, and at present poor rural local governments are generally lacking in sufficient transfers. Fiscal disparities are wide. For local areas, off-budget sources of revenue have proliferated, particularly since local governments are mainly responsible for providing basic education and health care for the rural population (Hussain and Stern 2008). Increasing the tax base continues to be one component of a strategy to improve public services, particularly in poor areas.

Livelihood

Not only has China's pattern of production adversely impacted the environment, but the pattern of consumption within the category of economic structure has also been detrimental. Consumption of food, retail products, automobiles, and water is rising. While green consumption and use of public transportation may combat some of the negative impacts of increasing consumption, they are currently insufficient to curb the worst environmental effects of rising demand for products and resources.

China is striving to continue to increase domestic consumption due to declining export demand and rising wages. The current policies, associated with the Twelfth Five Year Plan, are to increase urbanization focus on improving the living standards of individuals so that they can purchase manufactured goods and services domestically. Although this may have adverse effects on the environment, the policies would also allow individuals to purchase a larger quantity of goods, including important consumption items such as food and labor saving devices.

Green consumption

China has attempted to increase consumption of green products. China's environmental labeling program began in 1993 and was initiated by what is now the Ministry of Environmental Protection (Zhang 2011). The program was begun to promote green production and green consumption. There are certification standards for 85 categories of products. China also has created a Green Food product certification, with the China Green Food Development Centre founded in 1992 to oversee the implementation of this type of food production (Paull 2008). Green food production is based on reduced inputs and high standards. China's level of green consumption remains somewhat low but is increasing.

DOI: 10.1057/9781137358509

Water usage

Water usage is another aspect of consumption. There is high regional disparity in water usage in China (Zhou and Tol 2005). Water is used to different extents by different sectors, and is used more efficiently in some sectors and regions than in others. Part of this, too, is due to water price, which is particularly a factor with poor households, as wealthier households tend to use more water in general. China's water demand can be broken down into agricultural, domestic, industrial and environmental usage (Amarasinghe et al. 2005). Agriculture uses the most amount of water, at 70% of total use, followed by industry, at 21% of total use, then human and livestock use at 8%. These different sectors are associated with various reasons for water-use inefficiency. In the agricultural sector, this may be attributed to water leakage in irrigation systems and networks (Zhou and Tol 2005). In industry, this may occur to due to a large number of water intensive industries that may lack sufficient technology to curb water usage. Improvements in efficiency of water usage by the agricultural and industrial sectors in particular can reduce overall water consumption.

Transportation

China has a dense transportation infrastructure, which includes railways, expressways, deep water channels, and airport hubs (Zhou and Szyliowicz 2006). China's railway mileage was third in the world after the US and Russia in 2004, while highway length was fourth in the world, and airline passenger volume was third in the world. Efforts to enhance the railway network have resulted in creation of new railway lines as well as the introduction of high speed rail lines, while road mileage has doubled since the eighties (Britannica 2013b). Extensive highway construction has corresponded to a rise in the number of motor vehicles. Because of this, China consumes a large amount of energy, particularly oil.

Health

Heath is one of the main components of well-being. In order to improve overall well-being of its citizenry, China is currently attempting to address the issues involved in health care, including difficulties seeing a doctor and high-out-of-pocket expenses for medical care (Hu et al. 2008). China's price of hospital admission is currently twice as much

DOI: 10.1057/9781137358509

as the average annual income of the lowest 20% of households. Health insurance coverage is insufficient to cover patient costs.

China's main medical insurance schemes include the basic medical insurance scheme, which covers urban workers and covered about 28% of the urban population in 2006, the urban resident scheme, which was created to cover urban residents not serviced by the basic scheme, the rural cooperative medical system, which covered 87% of rural residents by 2007, and the medical assistance program, which covered very poor individuals (Hu et al. 2008). Policymakers are aware of the shortcomings of the health care system, and are attempting to address them.

China suffers from diseases that afflict both developed and developing countries, including AIDS, pulmonary tuberculosis, viral hepatitis, and dysentery. Table 2.7 contains a list of infectious diseases reported and the number of deaths caused by these diseases.

At the top of the list are viral hepatitis, pulmonary tuberculosis, syphilis, dysentery, and gonorrhea. Better health care might improve prevention and treatment for these diseases. Improvements in China's health care policy however cannot prevent the maladies brought about by the government mandated one-child policy. China's one child policy, which prevents urban residents from having more than one child, has resulted in an increase in the number of sex-selective abortions or female infanticide. About 1.1 million excess birth of males occurred in 2005 (Zhu, Lu and Hesketh 2009); about 336 million abortions and 222 million sterilizations have been carried out since 1971 (Ma 2013). The enforcement of the policy can be brutal, with forced abortions occurring if a woman gets pregnant and cannot pay the fine. As a result, China has the highest number of adult female suicides in the world, as women have lost their own reproductive rights. This is a policy that must be altered if the health of women and infants is to improve.

Science and technology

Improvements in science and technology, encouraged by policy, have aided economic production from both producer and consumer perspectives. Research and development has been geared toward production of innovative products, and telecommunications technology has been geared toward human and firm connectivity.

DOI: 10.1057/9781137358509

TABLE 2.7   *Ranking list of infectious diseases reported and number of deaths of class A and B (2010)*

Diseases		Death	
Diseases	Persons	Diseases	Persons
Viral Hepatitis	1317982	AIDS	7743
Pulmonary Tuberculosis	991350	Pulmonary Tuberculosis	3000
Syphilis	358534	Hydrophobia	2014
Dysentery	252248	Viral Hepatitis	884
Gonorrhea	105544	A(H1N1)Flu	147
Measles	38159	Hemorrhage Fever	118
Brucellosis	33772	Encephalitis B	92
Scarlet Fever	20876	Newborn Tetanus	86
AIDS	15982	Syphilis	69
Typhoid and Paratyphoid Fever	14041	Dysentery	36
Hemorrhage Fever	9526	Epidemic Encephalitis	33
Malaria	7389	Measles	27
A(H1N1)Flu	7123	Malaria	14
Schistosomiasis	4317	Leptospirosis	11
Encephalitis B	2541	Anthrax	6
		Typhoid and Paratyphoid Fever	3
Hydrophobia	2048	The Plague	2
Pertussis	1764	Gonorrhea	1
Newborn Tetanus	1057	Pertussis	1
Leptospirosis	677	Brucellosis	1
Epidemic Encephalitis	325	HpAI	1
Anthrax	289	Cholera	
Dengue Fever	223	Scarlet Fever	
Cholera	157	Dengue Fever	
The Plague	7	Schistosomiasis	
HpAI	1	SARS	
SARS		Poliomyelitis	
Poliomyelitis		Diphtheria	
Diphtheria			

Source: National Bureau of Statistics (2012).

China's research and development system was reformed in 1985 from a state-controlled to a market-oriented system (Kim and Mah 2009). However, until the mid-nineties, China was dependent upon imported technologies. To change this, in 1995, research and development activity was encouraged through tax incentives, establishment of science parks, and financial support for research and development activities. In 1999, partial tax deductions for research and development activities were implemented, along with tax exemptions for income from transfer of new technologies, preferential value added tax rates for software

DOI: 10.1057/9781137358509

products developed in China, and value added tax exemption for high tech exports. Since this period, the government has continued to encourage research and development activities. Currently, the Innovation Funds and Science and Technology Promotion Funds administered by the Chinese government encourage innovation in science and technology. These policies have enhanced China's research and development activities, and policies continue to strive to make China a country of innovation.

Telecommunications

The teledensity, or the number of main lines per 100 persons, has increased from 0.38% in 1978 to over 26% in 2008 (Fu and Mou 2010). This is a massive improvement. China's telecommunications industry was comprised solely of the state owned China Telecom until 1994, when the China United Telecommunications Corporation was created. China Telecom was later separated geographically into China Telecom and China Network Communications, while the wireless section of China Telecom was separated under China Mobile. Further restructuring was performed to facilitate the rollout of the 3G network. Currently, many individuals own cell phones and China is one of the top nations for cell phone usage.

Urban and rural culture

Urban and rural culture, as a facet of human and social well-being, faces some issues in sustainability. Although China has attempted to protect ethnic minorities, some of these groups continue to face high levels of discrimination. One can imagine that any type of culture that has developed in an unsustainable way, promoting creation of money over almost everything else, is not always successful in celebrating diversity.

China also has 56 officially recognized ethnic groups. The Han Chinese comprise the majority ethnic group at 92% of the population, while the other 55 ethnic groups are minorities. 44 of these groups occupy their own autonomous regions. Although some of China's ethnic groups are well integrated into the country's culture, some groups face difficulties, even violence, with the Han majority. Groups experiencing conflict with the majority include the Uighurs and the Tibetans (BBC 2009). Unrest in

DOI: 10.1057/9781137358509

Uighur regions has resulted in Han Chinese troop increases in the region and clashes between the ethnic groups. Peaceful demonstrations by Tibetans have been met with confrontation by Han Chinese authorities in Tibet. Due to these conflicts and even violence, some individuals possessing cultures different from the majority Han culture (especially Uighurs and Tibetans) have lower levels of well-being than the Han Chinese.

China has been more successful in protecting cultural heritage sites. China has signed several conventions for the protection of cultural heritage sites (Gruber 2007). These include the 1954 Convention for the Protection of Cultural Property in the Event of Armed Conflict, the 1972 UNESCO Convention for the Protection of the World Cultural and Natural Heritage, the 2001 UNESCO Convention on the Protection of the Underwater Cultural Heritage, the 1970 Convention on the Means of Prohibiting and Preventing the Illicit Import, Export and Transfer of Ownership of Cultural Property and the 1995 UNIDROIT Convention on Stolen or Illegally Exported Cultural Objects. These conventions cover monuments, historic buildings, and shipwrecks. Destruction of cultural heritage sites is not allowed, including in cases of war or economic development.

Well-being

Well-being has been impacted both by the economic structure and by the environment. High levels of inequality, environmental degradation, and insufficient spending on social services have resulted in lower levels of well-being among particular social groups (such as poorer Chinese) and in particular areas (such as rural and polluted regions).

China's environment directly affects individuals' well-being. A survey performed on 30 Chinese cities in 2003 found that high levels of atmospheric pollution, environmental disasters and traffic congestion negatively impact reported well-being while access to parkland and improved environmental conditions positively impact reported well-being, ceteris paribus (Smyth, Mishra and Qian 2008). This is as one would expect, since environmental degradation takes a toll on both physical and mental health while environmental improvements enhance these factors.

Economic income also impacts well-being. China has eliminated much of its worst poverty. Using China's national poverty line, rural poverty declined from 250 million in 1978 to 28.2 million in 2002 (Hu, Hu and Chang 2003). The percentage of China's population in food poverty

DOI: 10.1057/9781137358509

(consuming fewer than 1920 kcal per day per capita) declined from 30% in 1979–1981 to 17% in 1990–1992, and then to 11% in 1996–1998 (Ghosh 2010). Economic growth has been responsible for eliminating much of the poverty and improving citizens' lives in this respect.

However, despite poverty reduction, China has a large and growing amount of income inequality, which has, in many cases, had the opposite effect on well-being. China's Gini coefficient, cited by the World Bank (2012) was at 42.5, a very high level, in 2005. One main reason is that China has preserved institutions that have exacerbated inequality. One of the greatest barriers to equality has been posed by the hukou system, in which those living in rural and urban areas are separated. This system has made it difficult for rural residents to move to urban areas. Rural residents find it difficult to impossible to obtain an urban hukou, which would allow them to obtain a better job and to gain access to urban social services. Sometimes the rural-urban divide in education itself plays a role in being unable to obtain an urban hukou; rural residents without higher-level education are less likely to obtain an urban hukou than those who are highly educated.

Moving to urban areas does not necessarily combat inequality between rural and urban regions. Rural migrants to the cities generally obtain lower-paying jobs and face worse working conditions and lower job security (Guo and Cheng 2010). Much of this has to do with the difference in education between rural and urban residents. Urban poverty relief programs are provided only to urban hukou holders. Although the central government is attempting to address the rural-urban imbalances, the complexity of this issue, coupled with the way in which the problem is being addressed (to urbanize rural areas), makes us question whether these imbalances will be resolved any time in the near future.

Income and resulting well-being are also affected by the tax system. The Chinese tax system is comprised of consumption taxes, a value-added tax, income, property, and other taxes (IRET 2010). Revenues from consumption taxes are double the ratio of revenues from income taxes. China's progressive income tax system improves income inequality by focusing on taxing wealthier individuals (Piketty and Qian 2009). This has a positive impact on well being.

Much of China's spending on social services, however, is concentrated in urban areas. Health spending comprised 65 percent of all health spending and rural spending comprised 35 percent in 2004 (Li and

Piachaud 2006). Partial privatization of health care and fee-for-service care in rural areas has put health services out of reach for many rural residents (Lipton and Zhang 2007). Pension systems are available for urban residents as long as residents pay into them, but they are not available for rural residents.

The division of education, health and pension services into localized institutions, in rural versus urban areas, has led to a divergence in quality and availability of services. Urban residents have access and pay less relative to their incomes, while rural residents lack access and pay more relative to their incomes for services. The central government wishes to address these issues but here again, this will take time.

Governance

Crime and education policies have greatly impacted economic structure and well-being. While the elimination of crime is a positive occurrence, harsh punishment for crimes adversely impacts human well-being, particularly when innocent individuals are severely punished. Education policies have had more positive results than crime policies, and have been widely implemented. Over time, China has increased levels of education, leading to increased productivity at work, and enhancing human well-being as a result.

Crime

Since China opened up, criminal activity has increased and become internationalized. Some of the worst types of crime have included criminal investment in factories, through the corruption of public servants. Organized crime in the form of low- and mid-level government officials who take bribes from criminals and protect the criminals from being persecuted is of great concern (Chin and Godson 2006). The Chinese government is concerned about organized crime, particularly since it threatens the authority of the Communist Party. Coupled with more pedestrian crime such as rural gangster bullying and extortion activity, criminal activity has become increasingly challenging to pursue and eradicate.

To assist the persecution of criminal activity, China revised its Criminal Law in 1997 and its Criminal Procedural Law in 1996

DOI: 10.1057/9781137358509

(Broadhurst and Liu 2004). However, administrative punishment allows police to punish minor violations as they wish, and this aspect of Chinese law has been sharply criticized by outsiders. China is also known for executing the highest number of people in the world. Harsh punishment, even death, has garnered great attention from international human rights activists.

Corruption

Corruption is one type of criminal activity. China notoriously suffers from corruption, especially where there is social heterogeneity, regulation and resource abundance (Dong and Torgler 2012). The Communist Party has aimed to eliminate corruption, but it is difficult to detect, since local officials may use their discretion to decide which laws to enforce and which not to. Even if authorities know which laws were enforced and which were not, they do not ultimately know whether officials were collecting bribes or putting less priority on certain legal enforcements due to the presence of a hierarchically ranked set of mandates (Birney 2013). As long as officials uphold high priority mandates, their failure to uphold less pressing mandates may not receive notice.

In addition, government auditing of provincial accounting irregularities often reveals potential presence of corruption, but in provinces with high levels of corruption, rectification is often insufficiently undertaken (Liu and Lin 2012). Corruption has often resulted in the transfer of funds from development projects to the hands of officials. Reports on kleptocratic acts such as smuggling funds out of the country, seizure of peasant lands for enrichment of local officials, and misuse of funds for sex and real estate have colored the view of foreign businesses toward operating in the country, and have frequently enraged the citizenry, which has engaged in protests against these types of acts. Corruption continues to be on the list of issues to better address, but opacity in daily government operating activities makes the task difficult.

Education

Education is a public good and an important aspect of governance, since it can enhance or hamper the economic structure, as well as well-being. In the 1980s, China universalized primary school education, and in the

DOI: 10.1057/9781137358509

late eighties and nineties, universalized nine-year compulsory education (Xin and Kang 2012). There are several indicators that primary enrollment has expanded over the past 30 years, including the net enrollment rate of primary school age children, the advancement rate of primary school graduates, the gross junior middle school enrollment rate of youngsters aged 12 to 14, and the advancement rate of junior middle school graduates.

In Table 2.8, we can see the net enrollment ratio in primary schools and the promotion rate from primary school to junior secondary school, junior secondary school to senior secondary school, and from senior secondary school to higher education, between 1990 and 2010.

From Table 2.8, we can see that the general trend is an increase in the promotion rate from primary school to junior secondary school, junior secondary school to senior secondary school, and from senior secondary school to higher education. While net enrollment in primary schools has remained constant over this period, more and more students have been going on to obtain higher levels of education. This has very positive direct and indirect implications—it is directly good for children, and indirectly good for the economy and society, as educated citizens contribute more skilled labor and more sophisticated ideas.

In China, senior secondary school entrance examinations determine which track of senior secondary education students may attend (Brandenburg and Zhu 2007). Students attend senior secondary school for three years, and are then prepared for the National College Entrance Examination, which will determine which college the students may attend. Students who wish to enter the labor force at an earlier age are educated in technical schools with three to four year programs. Access to higher education has improved, such that in 1995, 5% of individuals aged 18 to 22 had access to higher education, while in 2007, 23% of college-age individuals had the same access (Zhao and Sheng 2008). In addition, annual college enrollment increased from one million in 1997 to 5.5 million in 2007. This is one of the most positive aspects of governance in China.

Participation

Public and international participation can enhance economic, environmental, and social well-being, promoting more inclusive policies. While China has become an active international participant, it has time and

DOI: 10.1057/9781137358509

TABLE 2.8 *Net enrollment ratio of primary schools and promotion rate of various schools*

Year	Net Enrollment Ratio of Primary Schools	Promotion Rate from Primary Schools to Junior Secondary Schools	Promotion Rate from Junior Secondary Schools to Senior Secondary Schools	Promotion Rate from Senior Secondary Schools to Higher Education
1990	97.8	74.6	40.6	27.3
1991	97.8	77.7	42.6	28.7
1992	97.2	79.7	43.6	34.9
1993	97.7	81.8	44.1	43.3
1994	98.4	86.6	47.8	46.7
1995	98.5	90.8	50.3	49.9
1996	98.8	92.6	49.8	51
1997	98.9	93.7	51.5	48.6
1998	98.9	94.3	50.7	46.1
1999	99.1	94.4	50	63.8
2000	99.1	94.9	51.2	73.2
2001	99.1	95.5	52.9	78.8
2002	98.6	97	58.3	83.5
2003	98.7	97.9	59.6	83.4
2004	98.9	98.1	63.8	82.5
2005	99.2	98.4	69.7	76.3
2006	99.3	100	75.7	75.1
2007	99.5	99.9	80.5	70.3
2008	99.5	99.7	82.1	72.7
2009	99.4	99.1	85.6	77.6
2010	99.7	98.7	87.5	83.3

Note: Data on promotion rate from junior secondary schools to senior secondary schools include those entering into secondary technical schools. Data on promotion rate from senior secondary schools to higher education refer to the ratio of new entrants into regular institutions of higher education (including regular classes of TV universities) to graduates of senior secondary schools.

Source: National Bureau of Statistics (2012).

again been called out for its lack of public participation and punishment of those who attempt to speak out against the government. In this way, human well-being in particular has been dampened, while economic, environmental, and social policies have remained in the control of the government.

International participation

For many years, China was isolated under Mao Zedong, but the country has clearly realized that it needed economic and political allies within

DOI: 10.1057/9781137358509

the framework of international organizations. China has been successful in becoming an internationally recognized participant in major regional and global organizations. These include the Asia Pacific Economic Cooperation, the United Nations, the G-20, the Asian Development Bank, the International Atomic Energy Agency, the World Trade Organization, and the World Health Organization, among others (CIA 2012). This shows that the rest of the world takes China seriously, and that China values its place as a global power.

Public participation

China's position in terms of public participation is far lower than that in terms of international participation. China has been attempting to increase public participation in the past 30 years. Starting in the eighties, the Chinese government strove to ensure that all laws and regulations were known so that the public could properly comply (Horsley 2009). Village self-governance and transparency policies of the eighties and nineties further promoted public participation. This was reinforced by former Communist Party General Secretary Jiang Zemin's praise for village self-governance and transparency. What is more, deliberative democracy experiments have been carried out at the local level, and online activism has developed as well.

However, China remains far behind democratic countries in allowing for public input on rulemaking and administrative decision-making (Horsley 2009). Freedom House rates China as 'Not Free' overall, nor in terms of internet or press participation. One example in which public participation was banned was construction of the Three Gorges Dam. Construction of the dam resulted in the relocation of 1.24 million residents, which was completed in 2008. Outcries against demolition of homes and forced relocation were dampened by the tightly controlled media, and protests were quashed by military troops (*Economist* 2002).

Construction of the dam had an adverse environmental impact as well, creating an inability to disperse pollutants effectively, an increase in algae blooms, and heightened soil erosion (Yang 2007). Watercraft and production accidents (about 12 per year) also continue to contribute to heavy pollution in the dam area (He et al. 2010). The environmental impact has thus been severe.

Further, China continues to repress its minority populations, especially the Uighur population, imposing violence to control them and cracking

DOI: 10.1057/9781137358509

down on peaceful demonstrations that protest this type of control. The government also arrests individuals who express any discontent with the state. The latest arrest of Ai Weiwei, a well-known artist who was openly critical about China's violation of human rights, made international news and exemplified China's status as a nation seriously lacking in freedom.

National resource accounting

National resource accounting can present clear indicators of environmental sustainability, which can greatly impact human well-being. However, national resource accounting, or "green" accounting in China, has been a hotly debated issue. China's Green GDP was calculated starting in 2004, but this ended when the government cancelled use of the metric in 2009 due to difficulties in estimating the market value of environmental conditions such as extinct species and health costs of pollution. Green GDP in China was calculated as the Integrated Environmental and Economic Accounting (Green GDP) by the State Environmental Protection Administration (SEPA) and the National Bureau of Statistics (NBS). It was also calculated in the National Environmental Pollution Loss Evaluation Survey by the SEPA, and in the forest resource accounting project undertaken by the NBS and the State Forestry Administration (SFA) (Wang, Jiang and Yu 2004). Ten regions were used to carry out pilot projects in green GDP in 2005 before the assessment was rolled out to 31 provinces and municipalities (Rauch and Chi 2010).

In addition, scholars have separately attempted to calculate variations of China's Green GDP. With reference to forests, Ying et al. (2011) find that the total stock value of forest lands increased between 1999 and 2003, and that in the value of forest services, the value of water conservation represented the largest contribution. Constructing an "ecoDP" indicator that includes eco-service domestic product, eco-disaster domestic product, ecoconstruction domestic product, eco-threat domestic product, and eco-protection cost input, Shi et al. (2012) find that between 1999 and 2008, eco-service domestic product rose by ¥0.11 trillion, eco-disaster domestic product increased by ¥1.03 trillion, eco-construction domestic product dropped by ¥9.8 billion, eco-threat domestic product rose by ¥114.4 billion, and ecoprotection cost input grew by ¥431.1 billion. Eco-service domestic product represents the positive value provided by natural ecosystem, eco-disaster domestic product represents the negative

DOI: 10.1057/9781137358509

value resulting from natural disasters, ecoconstruction domestic product represents the positive value provided by ecological and environmental construction, eco-threat domestic product represents the negative value resulting from human activities' stress on natural ecosystems, and eco-protection cost input the cost for ecological and environmental protection. Tabulation of China's Green GDP will now be left only to scholars and those outside of the state, as the government no longer performs this type of national resource accounting.

Property rights

Property rights policies in China have been implemented to control intellectual property rights and land ownership rights, but are generally considered insufficient to develop a healthy economic structure, a sustainable environment, and positive well-being. Laws are insufficient to protect individuals and organizations, and are not properly enforced.

Property rights in China constitute a serious issue from at least two perspectives. Literature on the subject has covered the topic of intellectual property rights as well as land ownership rights. Intellectual property rights have been a contentious issue between China and Western nations over the past two decades, and Chinese leaders have made commitments to crack down on piracy and other intellectual property rights infringements. Intellectual property rights have also been an issue for Chinese firms. Firms that have higher levels of intellectual property right protection, through government services for example, are more likely to invest in research and development (Lin, Lin and Song 2010). However, in many cases businesses are relatively unprotected, and it is well known that businesses must protect themselves against the threat of other firms stealing their ideas. Cases in which entire production facilities, not just individual products, were copied without punishment in China are testament to how far China has to go in protecting intellectual property rights (Keupp, Beckenbauer, and Gassman 2009).

Land ownership rights have also been frequently violated by government officials, as land takings increased in the late 1990s and early 2000s to make way for infrastructure of private corporations. The Rural Land Contracting Law of 2003 sought to reverse this trend by increasing the security of land use rights and enabling individuals to register appeals against violators of the law. Deininger and Jin (2009) find that this law

DOI: 10.1057/9781137358509

significantly reduced land takings where local leadership was democratically elected, and not where leadership was not democratically elected. This continues to be an issue to be addressed.

Energy self-sufficiency and international politics of energy markets

Energy policies impact economic structure, environmental sustainability, and human well-being. China previously pursued a policy of energy self-sufficiency, and is currently attempting to secure energy resources in countries with notorious political regimes. Attempts to secure energy resources have reassured investors in the economy, but have not necessarily promoted "green" attitudes toward energy or international approval of partnerships with questionable states. China's energy policies promote the status quo in terms of the economic structure, do not promote environmental sustainability, and are in line with autocratic attitudes toward the populace, reflecting a lack of support for human well-being.

For some time, an important goal for China was to be energy self-sufficient, and from the 1950s to 1970s, it was. As China's economy expanded, it lost energy self-sufficiency but increased its energy security, as relations with international neighbors improved (Zha 2006). China became a large exporter of coal and oil in exchange for industrial plant and technology from developed countries, and eventually began to import oil in 1983 as its energy demands rose further. Chinese oil companies began to go abroad in search of oil resources in 1993.

China has diversified oil sources in case a security threat arises. However, as the world becomes increasingly dependent on China for the production of goods, a politically motivated embargo has become more and more unlikely (Zha 2006). China generally pursues a policy of political diplomacy in order to secure energy resources.

At the same time, China has been rebuked by developed nations for engaging in energy trade with states that have poor ties to the West. The need to secure energy resources has led China to become friendly with Iran, Sudan, Uzbekistan and Venezuela (Zha 2006). This has been a concern for the West and particularly the United States. China's cooperation with Iran over oil has incensed the United States, which views Iran as a politically unfriendly state. What is more, China is currently building an oil pipeline linking it to Kazakhstan, which has awkward geopolitical

DOI: 10.1057/9781137358509

implications for China's relationship with Russia, a state through which the oil pipeline would have to pass in order to join the two countries.

China has also faced heavy criticism due to its heavy involvement in Africa. China has made partnerships with unpopular regimes in Africa in order to purchase minerals and energy sources. Although the share of imports from Africa is relatively small, it is rapidly increasing. What is more, China views sub-Saharan Africa as a commercial opportunity, particularly in resource-rich areas (Mol 2011).

Table 2.9 shows the total production and consumption of energy and to what extent coal, crude oil, natural gas, hydropower, nuclear power, and wind power comprise these categories, from 2006 to 2010.

From Table 2.9, we can clearly see that coal comprises the largest percentage of production and consumption, both of which are rising over this period. Production and consumption of the "greenest" energy category, hydropower, nuclear power, and wind power, are relatively low but slowly rising. Additional focus on green energy would help China turn away from its reliance on "dirty" energy.

Implication for the rest of the world

China's sustainable development practices have large implications for the rest of the world. China's environmental degradation has contributed substantially to global warming and overall reduction of natural resources and biodiversity. Although the country is trying to increase use of renewable energy resources and improve production efficiency, pollution of air and water continues to loom large as a problem in most of China's urban areas and many rural areas. China's pattern of development has not been sustainable and, although the current leadership strives to reach environmental sustainability, this is a massive undertaking that needs more attention.

China's record of human well-being, while performing well on an economic level, has not done well in terms of human rights and public participation, and income equality. The government is currently attempting to improve access to health care and other social services. To the extent that the rest of the world plays a role in sustaining China's poor record of human well-being—in terms of refraining from speaking out at a high political level against human rights abuses, and also in terms of maintaining income inequality by paying low wages through

DOI: 10.1057/9781137358509

TABLE 2.9 *Total production and consumption of energy and its composition*

	2006	2007	2008	2009	2010
Total Energy Production (10 000 tons of SCE)	232167	247279	260552	274619	296916
As Percentage of Total Energy Production (%):					
Coal	77.8	77.7	76.8	77.3	76.5
Crude Oil	11.3	10.8	10.5	9.9	9.8
Natural Gas	3.4	3.7	4.1	4.1	4.3
Hydro-power, Nuclear Power, Wind Power	7.5	7.8	8.6	8.7	9.4
Total Energy Consumption (10 000 tons of SCE)	258676	280508	291448	306647	324939
As Percentage of Total Energy Production (%):					
Coal	71.1	71.1	70.3	70.4	68
Crude Oil	19.3	18.8	18.3	17.9	19
Natural Gas	2.9	3.3	3.7	3.9	4.4
Hydro-power, Nuclear Power, Wind Power	6.7	6.8	7.7	7.8	8.6

Note: The coefficient for conversion of electric power into SCE (standard coal equivalent) is calculated on the basis of the data on average coal consumption in generating electric power in the same year.

Source: National Bureau of Statistics (2012).

Western multinationals based in China—the attitude toward China as a "cash cow" needs to be altered to take into account the human element. Although China's economic growth has benefited the rest of the world, much of it has been accomplished at the expense of its citizenry, and this must change.

DOI: 10.1057/9781137358509

3
Sustainable Development Programs and Experiments

Abstract: *Both Taiwan and China have recognized the need to improve sustainable development, and have increased the number of sustainable development programs and experiments that they are involved with. Although a comprehensive list is not provided, in this chapter, we list and discuss some of these.*

Hsu, Sara. *Lessons in Sustainable Development from China & Taiwan.* New York: Palgrave Macmillan, 2013. DOI: 10.1057/9781137358509.

DOI: 10.1057/9781137358509

Both Taiwan and China have recognized the need to improve sustainable development, and have increased the number of sustainable development programs and experiments that they are involved with. Although a comprehensive list is not provided, below, we list and discuss some of these.

Taiwan's sustainable development experiments

Taiwan has many successful sustainable development cases. These include:

1   The cleanup of Kaohsiung Harbor since its change from an industrial location to a tourist attraction. Kaohsiung has a master plan to clean up pollution, control floods, and improve recovery of the water environment (Yang 2012).

2   The cleanup of the Tamsui River (淡水河): Taipei City and New Taipei City launched the Taipei River Management Committee in 2011 to clean up the river and make it a tourist destination (Mo 2011).

3   Oil cleanup in Gaomei wetland area: A team of 50 soldiers and over 100 cleanup personnel was mobilized to clean up an oil spill dumped by unknown sources in Gaomei wetland area in central Taichung (*China Post* 2010). An investigation was carried out to identify the polluters.

4   Taiwan's Southern Taiwan Science Park has also been constructed as a model green high-tech park (National Council for Sustainable Development 2010). The park focuses on developing green industry through clean production and sustainable land use.

5   The town of Taicang in southern Jiangsu Province has provided a model for rural development, in collective economy, on the ground (Liu, Zhang and Zhang 2009). Taicang performs well in terms of both equity and environmental sustainability. The region is on the Yangtze River Delta Plain where land is arable and water abundant.

6   The Taiwan Sustainable Campus Programme (TSCP) was created in 2002 to convert the National Taiwan Normal University into a sustainable campus.

7   The Department of Water Quality Protection within the Environmental Protection Administration has cleaned up Keelung City's Tianliao River, Taipei County's Jhong-Gang Drainage Channel, Kaohsiung County's Fengshan River, Pingtung County's

DOI: 10.1057/9781137358509

Wannien River, and Taichung City's Liuchuan River by improving household wastewater treatment (EPA of Taiwan 2010).

8 The Water Resources Bureau in Taipei County has focused on remediation of the Danshui River, building on-site treatment systems and installing a water purification plant while listening to community residents and using environment-friendly materials (EPA of Taiwan 2010).

9 The Leave No Trace Movement has worked toward setting up a national trail system. The foundation for trail infrastructure was created and mapped, and 11 teacher training sessions trained 200 people on the Leave No Trace concept (EPA of Taiwan 2008).

10 The Bureau of Energy provided 541 subsidies for photovoltaic installations between 2000 and 2008. Many of these were installed onto elementary schools as part of the "Sunshine Campus" plan. Others were implemented into agricultural projects and fulfilled as Council of Agriculture applications (EPA of Taiwan 2008).

11 The Council of Indigenous Peoples drafted regulations to utilize traditional knowledge in protecting biodiversity. Expert surveyors from each tribe contributed to the draft, which was sent on to the Executive Yuan. Information was compiled into a biological database in the Academia Sinica Taiwan Biodiversity Information Facility (EPA of Taiwan 2008).

12 Several health risk databases were created, including the "Compilation of Exposure Factors of the Average Person in Taiwan," the "Study on Health Status of Residents Near Dense Installations of Mobile Phone Base Stations," and the "Non-ionizing Radiation Health Risk Communication Plan" (EPA of Taiwan 2008).

13 The Center for Disease Control has worked to set up an efficient infectious disease surveillance system to control epidemics and infectious disease transmission (EPA of Taiwan 2008).

14 The Dongsha Marine National Park was created to protect Dongsha Atoll. This has increased the area covered by live coral as well as the distribution of native plants, and provides a resource for marine preservation (EPA of Taiwan 2008).

From these examples, we can see that Taiwan is clearly striving toward sustainable development. Every year, new sustainable development programs arise and some of these are given awards by the government.

DOI: 10.1057/9781137358509

Not only is Taiwan in the process of cleaning up their environment, but they are also implementing proper monitoring for indicators of sustainable development, experimenting with sustainable campuses, and adding elements of "green" energy. Taiwan is proactive in sustainable development.

China's sustainable development experiments

China's sustainable development successes include:

1 The Western Development Program, begun in 1999, was aimed toward the promotion of growth in the impoverished western regions of China. The program aimed to provide job opportunities through incentives provided to companies to move to the regions, and through the construction of infrastructure. The program was also set up to enhance environmental protection in the region and to promote education.

2 China's Eco-Communities (*Shengtai Qu*) were developed in 1995 in order to experiment with sustainability practices (Liu 2008). These communities were to set up their own Environmental Protection Bureau and to ensure that environmental protection policies were implemented. Although mainly located in wealthier provinces, of the 42 eco-communities, 25 were set up in the less developed region of their provinces.

3 Xiamen's local government has set up an integrated coastal zone management program initially under a 5-year demonstration program starting in 1994, and extended through the present (Lau 2003). To do this, the Xiamen government included scientists, economists, legal experts, engineers, and urban planners to provide expertise to policymakers, helping to create scientific tools for coastal management policy.

4 Several provinces have made use of (or are in the process of installing) wind turbines for energy generation, including Gansu, Fujian, Heilongjiang, Jiangsu, Inner Mongolia, Xinjiang, Jilin, Hebei, Shandong, Ningxia, and Liaoning provinces.

Although China is making some efforts toward sustainable development, it lags far behind Taiwan and can learn much from the latter country. China still struggles with enforcing environmental regulations and

DOI: 10.1057/9781137358509

many industries and regions continue to be highly polluting. The nation has made far less progress in cleaning up its environment. Implementing sustainable development programs at the pace at which Taiwan implements its own programs right now seems impossible as China has a host of concerns to address. The country is much larger geographically and less equitable in terms of economic structure. These factors have greatly impacted the environment and well-being. From the outset, then, China has had more issues to cope with than Taiwan.

DOI: 10.1057/9781137358509

4
Directions for Future Study

Abstract: *China and Taiwan's large increase in industrialization and urbanization over the years has produced economic growth without producing an equal amount of sustainable development. In this chapter, we discuss policy recommendations for China and Taiwan, how the countries can learn from one another, and how a partnership between the two countries might assist in the process of sustainable development.*

Hsu, Sara. *Lessons in Sustainable Development from China & Taiwan.* New York: Palgrave Macmillan, 2013. DOI: 10.1057/9781137358509.

China and Taiwan's large increase in industrialization and urbanization over the years has produced economic growth without producing an equal amount of sustainable development. Much of this is due to the industrial nature of growth, involving extensive use of inefficient factories and polluting materials. In some ways, these nations are at a disadvantage. Compared to less developed countries like Indonesia, which currently has many sustainable development experiments under way, China and Taiwan produce using advanced methods that generate environmental pollution. Reversal of the economic structure is difficult as it has already become institutionalized. However, as we can see from the section above on sustainable development experiments, amelioration of adverse effects of economic development can be cultivated.

Policy recommendations for China

Although there is disagreement among policymakers at the international level as to how environmental sustainability should be regulated, with Chinese leaders making the case that China should be allowed to emit as much carbon per capita as the US (c.f. Parks and Roberts 2008), at a practical level, carbon reduction must occur in all nations to postpone rapidly approaching temperature increases.

Because most of the industrialization in China has occurred in cities, some of the worst environmental degradation, as well as the greatest wealth, are located in cities. Cities are rapidly expanding, and sustainable development should seek to prevent worst practices and enhance best practices for healthy living in cities and rural areas. The focus, then, is on a dually aimed policy, of improving opportunity for income generation in rural areas, and on reducing environmental degradation and inequality in urban areas.[1]

China is turning gradually toward implementing greener forms of energy and other types of sustainable development activities, but one should strongly question whether the rate at which China is improving its environmental situation is fast enough to offset an irreversible climate change. Further research might examine whether the shift in focus away from partial market share by multinational corporations and toward promotion of smaller domestic companies would also potentially have the impact of improving environmental circumstances; if small and medium sized enterprises can be more easily regulated than multinationals, for example, or are more likely to pay taxes on pollution, their promotion

DOI: 10.1057/9781137358509

in place of (rather than alongside) multinational corporations might be desirable. Strong consideration of finding ways to preserve some rural ways of life, rather than imposing growth through incentives to large corporations, is encouraged in order to prevent further environmental degradation.

China lacks the level and universality of social spending that Taiwan maintains, and its current social spending structure reinforces income inequalities. China's progressive taxation system has a positive impact on enhancing income equality, but additional gains from taxation by the central government must be channeled into rural social services if inequality is to be alleviated. Attempts by the Western Development Program to bring growth to the Western region have helped reduce poverty somewhat, and the program has a positive impact on enhancement of equality. More can be done, however, in regions where most of the population is concentrated; over 60 percent of the population lives outside the Western provinces of focus in the Western Development Program.

The government is currently attempting to rebalance the urban and rural sectors and to enhance rural social services. It is very much aware of the problems, but with an agenda teeming with important issues, this rebalancing may take some time. China is currently addressing inequality through an urbanization policy that focuses on urbanizing rural residents. Whether this is the best policy to improve the lot of rural residents in an environmentally sustainable manner remains a looming question.

What can China learn from Taiwan in terms of sustainable development?

Taiwan's inclusive pattern of growth, with a focus on small and medium enterprises rather than large corporations, is instructive. Much of China's GDP is produced by small and medium enterprises (SMEs), although China has also welcomed large corporations, particularly multinationals. Further examination of the impact of a reduction in market share of multinationals would be beneficial in determining whether the country could improve its income equality status. An improved policy environment for SMEs may be one key to reducing the high degree of inequality.

In addition, Taiwan's Environmental Protection Bureau is an effective institution that China may seek to emulate. Taiwan's EPA has been successful in setting up a sustainable development program, as well as

DOI: 10.1057/9781137358509

in coordinating the implementation of the program across various government agencies. The bureau is also highly responsive in cleaning up environmental damage and in assisting enforcement of environmental laws. China's environmental institutions lack the authority to enforce environmental laws and embark on their own full-fledged sustainable development programs. Empowering China's environmental institutions is essential to improving the country's environment.

Some aspects of sustainable development such as social protection and health are progressing in China, while others, such as human rights/public participation and sustainable production lag behind. While China clearly has an agenda full of issues to address, some of its sustainable development indicators will continue to lag because they are embedded in its economic and political institutions. Lack of democracy and a heavy emphasis on economic growth ensure that China will not "catch up" to Taiwan in terms of sustainable development. Changing these fundamental factors would mark distinct progress toward sustainable development.

Policy recommendations for Taiwan

Taiwan has a high level of human well-being. Taiwan's equality situation is good, and additional social spending can further improve the situation of the poorest individuals. Taiwan's main focus, however, must be on improving the environment. The country's population pressures are immense and a great source of pollution. Renewable energy has not yet been implemented and is in need of becoming a policy focus. The recent implementation of laws taxing pollution will hopefully curb some of the worst pollution so that the island-state can move forward in promoting sustainable development. No doubt the country is attempting to enhance sustainable development, but it faces difficulties due to existing pollution, excessive number of motor vehicles, and constrained geography from which to produce "green" energy.

What can Taiwan learn from China in terms of sustainable development?

China's way of implementing sustainable development experiments can provide a model for other countries that are attempting to change their

DOI: 10.1057/9781137358509

development regimes. Where potential ramifications of new practices are unknown and/or potentially costly, setting up better practices in smaller areas as "experiments" can be beneficial. Taiwan has set up wind farms across the western region, but expanding the practice of setting up experiments within regions may help Taiwan in further implementing greener energy practices. China is truly the master of the experimental approach, and has used this for a variety of political and economic policies, adjusting the experiments to best suit the reality on the ground. This is something that Taiwan can emulate.

Partnership between China and Taiwan

A partnership between China and Taiwan in exploring alternative energy might be particularly beneficial since there is a pressing need for both countries to quickly develop and implement such technologies. Transitioning to greener energy is critical to prevent accelerating pollution, but it is expensive to implement (as compared to implementing nothing new), demanding of new skills, and representative of an environmental "regime change" to which locals firms and individuals are unaccustomed. Friendly competition and cooperation between the two nations in increasing usage of green energy might be highly beneficial.

In addition, if the technology to store and transmit "green" energy is improved, the two countries can benefit from producing green energy for one another. China's vast Western plains are ideal for generating wind power and transmission to eastern regions, and Taiwan would help to alleviate some reliance on "dirty" energy sources. Taiwan may be able to produce and transmit some ocean energy in return. Both countries can exchange technology research with one another to enhance sustainable development together.

Notes

1   A sustainable development model laid out by Wei, Tsai, Fan and Zeng (2004) suggests that in order to reduce pollution and increase personal income overall and as part of the agricultural, industry, and service sectors, population growth controls are relevant, as are a shift in employment toward

DOI: 10.1057/9781137358509

the service sector, creation of employment opportunities, and saving in municipal water consumption. The authors also find that degradation of resources will impose a severe bottleneck in the future, and that improving environmental indicators and environmental investments are critical.

DOI: 10.1057/9781137358509

Conclusion

Abstract: *In the conclusion, we summarize the status of sustainable development in China and Taiwan and revisit some policy recommendations.*

Hsu, Sara. *Lessons in Sustainable Development from China & Taiwan.* New York: Palgrave Macmillan, 2013.
DOI: 10.1057/9781137358509.

▶

In this book, we have discussed many facets of sustainable development in China and Taiwan. While Taiwan's equality situation is much more favorable than that of China, it harbors many of the same environmental challenges to sustainable development that China faces. Legislation in and of itself has not eradicated environmental ills in Taiwan, and the pressing need for proactive policies combating climate change underscores the importance of enhancing the regulatory and monitoring structure for polluters, and of finding viable sources of alternative energy. Taiwan struggles with a relatively small and mountainous land mass and a high population density in its effort to implement greener energy solutions and practices, but these barriers must be overcome for sustainable development to be truly brought about.

China's equality situation has been widely commented on and is stark, particularly between the rural and urban regions, which has resulted from its urban, coastal-biased pattern of development. Implementation of social services in rural areas is essential, as is creating sustainable growth in rural regions. A multipronged sustainable development policy in China, as has been implemented in Taiwan, may be more effective in addressing the severe environmental and inequality issues in the country. This will be costly and will take time, but it will improve the lives of billions of people as a result.

Some partnership between Taiwan and China in exploring more sustainable energy sources could help both countries to quickly improve their environmental statuses. Although we cannot recommend a partnership in enhancing social aspects of sustainable development, due to the different approaches embedded in political regimes of both nations, a partnership to accelerate and enhance energy security and technology research may be highly desirable.

DOI: 10.1057/9781137358509

References

Amarasinghe, Upali A., Mark Giordano, Yongsong Liao, and Zhongping Shu. 2005. "Water Supply, Water Demand and Agricultural Water Scarcity in China: A Basin Approach." *International Commission on Irrigation and Drainage Country Policy Support Programme Report 11.*

Ashby, W. Ross. 1956. *Introduction to Cybernetics.* London: Chapman and Hall.

Ashton, Weslynne, Andres Luque, and John R. Ehrenfeld. 2002. *Best Practices in Cleaner Production Promotion and Implementation for Smaller Enterprises.* Report prepared for Multilateral Investment Fund and Interamerican Development Bank, Washington, DC.

Bartley, Tim and Lu Zhang. 2012. "Opening the 'Black Box': Transnational Private Certification of Labor Standards in China." Indiana University. RCCPB Working Paper #18.

BBC. 2009. "China's Main Ethnic Minorities." July 6. http://news.bbc.co.uk/2/hi/asia-pacific/8136043.stm

BBC. 2011. "Taiwan Endangered Species Focus of New Awareness." February 1. http://www.bbc.co.uk/news/world-asia-pacific-12208493

Bhalla, Ajit S., Shujie Yao and Zongyi Zhang. 2003. "Causes of Inequalities in China, 1952 to 1999." *Journal of International Development* 15: 939–955.

Birney, Mayling. 2013. "Decentralization and Veiled Corruption under China's 'Rule of Mandates.'" *World Development.* Article in Press.

Blaikie, Piers and Harold Brookfield. 1987. *Land Degradation and Society.* London: Methuen.

DOI: 10.1057/9781137358509

Bossel, Hartmut. 1999. *Indicators for Sustainable Development: Theory, Method, Applications: A Report to the Balaton Group.* Winnepeg, Canada: International Institute for Sustainable Development.

Brandenburg, Uwe and Jiani Zhu. 2007. "Higher Education in China in the Light of Massification and Demographic Change: Lessons to be learned for Germany." Centrum für Hochschulentwicklung gGmbH Working Paper 97, October.

Britannica. 2013a. "Taiwan." http://www.britannica.com/EBchecked /topic/580902/Taiwan/30006/Religions Accessed July 2.

Britannica. 2013b. "China." http://www.britannica.com/EBchecked /topic/111803/China. Accessed July 19.

Broadhurst, Roderic G. and Jianhong Liu. 2004. "Introduction: Crime, Law and Criminology In China." *The Australian and New Zealand Journal of Criminology* 37(Supplement): 1–12.

Busby, Joshua W. 2010. *China and Climate Change: A Strategy for U.S. Engagement.* Washington, D.C., Resources for the Future Report, November.

Chan, David Yih-Liang, Kuang-Han Yang, Chung-Hsuan Hsu, Min-Hsien Chien, Gui-Bing Hong. 2007. "Current Situation of Energy Conservation in High Energy-Consuming Industries in Taiwan." *Energy Policy* 35: 202–209.

Chan, Hou-sheng. 2008. "The Development of Social Welfare Policy in Taiwan: Welfare Debates between the Left and the Right." Paper delivered at Department of Sociology, Doshisha University, Kyoto, January 25.

Chan, Kam Wing. 2010. "A China Paradox: Migrant Labor Shortage amidst Rural Labor Supply Abundance." *Eurasian Geography and Economics,* 51(4): 513–530.

Chang, Gene H. 2002. "The Cause and Cure of China's Widening Income Disparity." *China Economic Review* 13: 335–340.

Chang, Ssu-Li. 2012. *An Overview of Energy Policy and Usage in Taiwan. In Wakefield Taiwan's Energy Conundrum.* Woodrow Wilson International Center for Scholars Special Report No. 146.

Chang, Tzi-Chin, Jiun-Horng Tsai, Yung-Chen Yao and Wen-Shing Chang. 2010. "Effectiveness of the Air Pollution Fee Strategy on Air Quality Improvement in Taiwan During 1995–2006." *Journal of Environmental Engineering Management* 20(1): 19–26.

Chen, Chun-Shuo and Terrence A. Maxwell. 2007. "The Dynamics of Bilateral Intellectual Property Negotiations: Taiwan and the United States." *Government Information Quarterly* 24: 666–687.

DOI: 10.1057/9781137358509

Chen, Falin, Shyi-Min Lu, Kuo-Tung Tseng, Si-Chen Lee, Eric Wang. 2010. "Assessment of Renewable Energy Reserves in Taiwan." *Renewable and Sustainable Energy Reviews* 14: 2511–2528.

Chen, Home-Ming and Shang-Lien Lo. 2010. "Economic Analyses for Optimizing the Construction of Separate Sewer in a Hybrid Sewer System." *Water Science and Technology* 62(11): 2536–2542.

Chen, Jian. 1996. "Regional Income Inequality and Economic Growth in China." *Journal of Comparative Economics* 22: 141–164.

Chen, Meei-Shia and Chang-Ling Huang. 1997. "Industrial Workers' Health and Environmental Pollution Under the New International Division of Labor: The Taiwan Experience." *American Journal of Public Health* 87(7): 1223–1231.

Chen, Ting-An. 2008. "Introduction to Taiwan's Taxation System: Major Problems and Reform." East Asian Tax Forum. http://www.ipp.hit-u.ac.jp/EastAsianTaxForum/PDF/chen-Introduction%20to%20Taiwan's%20Taxation%20System.pdf.

Chen, Wen. 2007. "Economic Growth and the Environment in China——An Empirical Test of the Environmental Kuznets Curve Using Provincial Panel Data." Paper prepared for the "Annual Conference on Development and Change" in Cape Town.

Chen, Zhao and Ming Lu. 2008. "Is China Sacrificing Growth when Balancing Interregional and Urban-Rural Development?" Chapter 15 in Huang, Yukon, Magnoli Bocchi, Alessandro (Eds.), *Reshaping Economic Geography in East Asia*, Washington: World Bank.

Cheng, Tsung-Mei. 2003. "Taiwan's New National Health Insurance Program: Genesis and Experience So Far." *Health Affairs*, 22(3): 61–76.

Chi, Gou-Chung. 2010. "Materials for Clean Energy Production and CO_2 Reduction." Seminar Presentation, National Chiao Tung University, www.phys.nchu.edu.tw/files_news/seminar2010100801.ppt.

Chiang, Min-Chin. 2010. "The Hallway of Memory: A Case Study on the Diversified Interpretation of Cultural Heritage in Taiwan." Chapter in *Becoming Taiwan: From Colonialism to Democracy*, eds. Ann Heylen and Scott Sommers. Gottingen, Germany: Hubert and Co.

Chiao, Cing-Kae. 2008. "Employment Discrimination in Taiwan." The Japan Institute for Labour Policy and Training Working Paper, http://web.jil.go.jp/english/events_and_information/documents/cllso8_chiao.pdf.

DOI: 10.1057/9781137358509

Chiau, Wen-Yen. 2005. "Changes in the Marine Pollution Management System in Response to the Amorgos Oil Spill in Taiwan." *Marine Pollution Bulletin* 51: 1041–1047.

Chin, Ko-lin. 2003. *Heijin: Organized Crime, Business and Politics in Taiwan.* New York: ME Sharpe.

Chin, Ko-lin and Roy Goodson. 2006. "Organized Crime and the Political-Criminal Nexus in China." *Trends in Organized Crime* 9(3): 5–44.

China Post. 2010. "Gaomei Suffers from Serious Oil Pollution." *China Post,* July 1.

China UN Mission. 2012. *China's National Report on Sustainable Development.* http://www.china-un.org/eng/zt/sdreng/P020120608816970051133.pdf

Choucri, Nazli. 1995. *Global Accord: Environmental Challenges and International Responses.* Cambridge: MIT Press.

Chuang, Ming Chih and Hwong Wen Ma. 2013. "An Assessment of Taiwan's Energy Policy Using Multi-Dimensional Energy Security Indicators." *Renewable and Sustainable Energy Reviews* 17: 301–311.

CIA. 2012. "China." World Factbook, July 31.

Council of Agriculture Forestry Bureau. 2012. "Biodiversity." http://www.forest.gov.tw/ct.asp?xItem=21503&ctNode=1887&mp=3 Accessed August 1.

Council for Economic Planning and Development, Executive Yuan. 2004. *Taiwan Agenda 21: Vision and Strategies for National Sustainable Development.* Taipei: Council for Economic Planning and Development.

Dasgupta, Partha. 1995. "Population, Poverty and the Local Environment." *Scientific American* 272(2): 26–31.

Dean, Judith M., Mary E. Lovely, and Hua Wang. 2009. "Are Foreign Investors Attracted to Weak Environmental Regulations? Evaluating the Evidence from China." *Journal of Development Economics* 90: 1–13.

Deininger, Klaus and Songqing Jin. 2009. "Securing Property Rights in Transition: Lessons from Implementation of China's Rural Land Contracting Law." *Journal of Economic Behavior and Organization* 70:22–38.

Demurger, Sylvie. 2001. "Infrastructure Development and Economic Growth: An Explanation for 'Regional Disparities in China?" *Journal of Comparative Economics* 29(1):95–117

Démurger, Sylvie, Yuanzhao Hou, and Weiyong Yang. 2009. "Forest Management Policies and Resource Balance in China: An

DOI: 10.1057/9781137358509

Assessment of the Current Situation." *The Journal of Environment Development* 18(1): 17–41.

Diao, Zeng, Tam and Tam 2009. "EKC Analysis for Studying Economic Growth and Environmental Quality: A Case Study in China." *Journal of Cleaner Production* 17(5): 541–548.

Dong, Bin and Benno Torgler. 2012. "Causes of Corruption: Evidence from China." *China Economic Review*-Article in Press.

Economist. 2002. "Dam Shame." *Economist,* July 4.

Environmental Protection Agency (EPA). 2011. "Taiwan Environmental Law Library." http://law.epa.gov.tw/en/laws/278076101.html.

EPA of Taiwan. 2008. *Annual Report on National Sustainable Development.* http://nsdn.epa.gov.tw/en/PRINT/97Annual.pdf

EPA of Taiwan. 2012a. *New National Sustainable Development Indicators.* http://nsdn.epa.gov.tw/en/TSDI/New%20Indicators.pdf

EPA of Taiwan, 2012b. http://www.epa.gov.tw/en/epashow. aspx?list=102&path=135&guid=c4b6adof-13e5-4259-be98-8356037dc862&lang=en-us

EPA of Taiwan, 2012c. Annual Report of Water Quality. http://wq.epa.gov. tw/WQEPA/Code/Report/ReportShow.aspx?ID=24&Languages=en

Fan, Chunliang. 2009. "China's Technology Policies Related to Sustainable Environment." *AIP Conference Proceedings* 1157: 93–100.

Feng, Fong-Long. 1997. "Environmental Assessment for Agricultural Development in Taiwan." http://fm4sem.nchu.edu.tw/%E5%B7%B2% E7%99%BC%E8%A1%A8%E8%AB%96%E6%96%87PDF%E6%AA%9 4/51-Eia-crpt.pdf

Freedom House. 2013. "Taiwan." Freedom in the World 2012. http: //www.freedomhouse.org/report/freedom-world/2012/taiwan

Fu, Hanlong and Yi Mou. 2010. "An Assessment of the 2008 Telecommunications Restructuring in China." *Telecommunications Policy* 34: 649–658.

Fujita, Masahisa and Paul Krugman. 2003. "The New Economic Geography: Past, Present and the Future." *Regional Science* 83(1): 139–164.

Gao, Pat. 2009. "Facing a Thirstier Future." *Taiwan Review,* January 2. http://taiwanreview.nat.gov.tw/ct.asp?xItem=47603&CtNode=119

Geiger, Michael. 2008. "Instruments of Monetary Policy in China and Their Effectiveness: 1994–2006." UNCTAD Discussion Paper 187.

Gemmer, Marco, Andreas Wilkes, and Lucie M. Vaucel. 2011. "Governing Climate Change Adaptation in the EU and China: An

DOI: 10.1057/9781137358509

Analysis of Formal Institutions." *Advances in Climate Change Research* 2(1): 1–11.

Gleick, Peter. 2010. "China and Water," Chapter in *The World's Water,* ed. Peter Gleick.. Washington, DC: Island Press.

Ghosh, Jayati. 2010. "Poverty Reduction in China and India: Policy Implications of Recent Trends." UN DESA Working Paper No. 92.

Gruber, Stefan. 2007. "Protecting China's Cultural Heritage Sites in Times of Rapid Change: Current Developments, Practice and Law." *Asia Pacific Journal of Environmental Law* 10(3–4): 253–301.

Guo, Fei and Zhiming Cheng. 2010. "Labor Market Disparity, Poverty, and Inequality in Urban China." *China Perspectives* 4: 16–31.

Hallding, Karl, Guoyi Han and Marie Olsson. 2009. *A Balancing Act: China's Role in Climate Change.* Sweden: The Commission on Sustainable Development.

He, Qiang, Shujuan Peng, Jun Zhai, Haiwen Xiao. 2010. "Development and Application of a Water Pollution Emergency Response System for the Three Gorges Reservoir in the Yangtze River, China." *Journal of Environmental Sciences* 23(4) 595–600.

Ho, Karl, Harold D. Clarke, Li-Khan Chen, Dennis Lu-Chung Weng. 2013. "Valence Politics and Electoral Choice in a New Democracy: The Case of Taiwan." *Electoral Studies-* Article in Press.

Horsley, Jamie P. 2009. "Public Participation in the People's Republic: Developing a More Participatory Governance Model in China." http://www.law.yale.edu/documents/pdf/Intellectual_Life/ CL-PP-PP_in_the__PRC_FINAL_91609.pdf

Hou, Yu and Tian-zhu Zhang. 2009. "Evaluation of Major Polluting Accidents in China—Results and Perspectives." *Journal of Hazardous Materials* 168: 670–673.

Hsu, Jinn-Yuh and Lu-Lin Cheng. 2002. "Revisiting Economic Development in Post-war Taiwan: The Dynamic Process of Geographical Industrialization." *Regional Studies,* 36(8): 897–908.

Hsu, Kuo-Cheng and Tsung-Yu Lai. 2007. "Institutions and Institutional Changes: the Case of Taiwan's Non-Urban Land Development System." Asian Real Estate Society Conference Paper.

Hsu, Minna J., Selvaraj, G. Agoramoorthy. 2006. "Taiwan's Industrial Heavy Metal Pollution Threatens Terrestrial Biota." *Environmental Pollution* 143: 327–334.

Hu, Angang, Linlin Hu and Zhixiao Chang. 2003. "China's Economic Growth and Poverty Reduction (1978–2002)." IMF Conference "A

DOI: 10.1057/9781137358509

Tale of Two Giants: India's and China's Experience with Reform and Growth," November 14–16, New Delhi.

Hu, Jin-Li, Mon-Chi Lio, Fang-Yu Yeh, Cheng-Hsun Lin. 2011. "Environment-Adjusted Regional Energy Efficiency in Taiwan." *Applied Energy* 88: 2893–2899.

Hu, Robert Yie-Zu. 2011. *"Current Status of Energy Efficiency Policies and Measures in Taiwan." 2011 U.S.-Taiwan Clean Energy Forum Presentation.*

Hu, Shanlian, Shenglan Tang, Yuanli Liu, Yuxin Zhao, Maria-Luisa Escobar, David de Ferranti. 2008. *Reform of How Health Care is Paid for in China: Challenges and Opportunities.* Publication in Series, Health System Reform in China, Brookings Institution.

Huang, Shu-Li, Szu-Hua Wang, and Budd, William W. 2009. "Sprawl in Taipei's Peri-Urban Zone: Responses to Spatial Planning and Implications for Adapting Global Environmental Change." *Landscape and Urban Planning* 90: 20–32.

Hubacek, Klaus, Dabo Guan, John Barrett, and Thomas Wiedmann. 2009. "Environmental Implications of Urbanization and Lifestyle Change in China: Ecological and Water Footprints." *Journal of Cleaner Production* 17(14): 1241–1248.

Hubacek, Klaus,Kuishuang Feng, and Bin Chen. 2012. "Changing Lifestyles Towards a Low Carbon Economy: An IPAT Analysis for China." *Energies* 5, 22–31.

Hung, Li-Ju, Shang-Shyue Tsai, Pei-Shih Chen, Ya-Hui Yang, Saou-Hsing Liou, Trong-Neng Wu, Chun-Yuh Yang. 2012. "Traffic Air Pollution and Risk of Death from Breast Cancer in Taiwan: Fine Particulate Matter (PM2.5) as a Proxy Marker." *Aerosol and Air Quality Research* 12: 275–282.

Hung, Shih-Chang. 2004. *Taiwan's Hsinchu Science-Based Industrial Park, evolution (1/2).* National Science Council Research Project Midterm Progress Report.

Hussain, Athar and Nicholas Stern. 2008. "Public Finances, the Role of the State, and Economic Transformation, 1978–2020." Chapter in *Public Finance in China: Reform and Growth for a Harmonious Society,* eds Jiwei Lou and Shuilin Wang. Washington, DC: World Bank.

Inglesby, Tom, Anita Cicero, Jennifer Nuzzo, Amesh Adalja, Eric Toner, Kunal Rambhia, and Ryan Morhard. 2012. *Report on Taiwan's Public Health Emergency Preparedness Programs 10 Years after SARS.* UPMC Center for Health Security Report.

DOI: 10.1057/9781137358509

Institute for Research on the Economics of Taxation (IRET). 2010. "The Tax System of China." IRET Policy Bulletin No. 94.

ICFTU. 2006. *Internationally Recognized Core Labor Standards in Chinese Taipei*. Report for the WTO General Council Review of the Trade Policies in Chinese Taipei. Geneva, June 20 and 22.

Jänicke, Martin Harald Mönch, Thomas Ranneberg, Udo E. Simonis. 1989. "Economic Structure and Environmental Impacts: East-West Comparisons." *Environmentalist* 9(3): 171–183.

Jao, Chih-Chien. 2000. "The Green Accounting in Taiwan." International Symposium on Indicators of Sustainable Development. http://mdgs.un.org/unsd/envAccounting/ceea/archive/Framework /Green_Accounting_Taiwan.PDF

Johnson, Ian. 2011. "China Outlines Cuts in Carbon Emissions." *New York Times,* November 22.

Jones, Derek C. Cheng LI, and Ann L. Owen. 2003. "Growth and Regional Inequality in China during the Reform Era." *China Economic Review* 14: 186–200.

Kanbur, Ravi and Xiaobo Zhang. 1999. "Which Regional Inequality? The Evolution of Rural–Urban and Inland–Coastal Inequality in China from 1983 to 1995." *Journal of Comparative Economics* 27(4): 686–701.

Kanbur, Ravi and Xiaobo Zhang. 2004. "Fifty Years of Regional Inequality in China A Journey through Central Planning, Reform, and Openness." UN Wider Research Paper No. 2004/50.

Kao, Cheng-Yan. 2008. "Development of Biomass Energy in Taiwan: Current Status and Prospects." Paper delivered at Conference on Energy, Environment, Ecosystems, and Sustainable Development, June 11–13, Algarve, Portugal.

Kaye, Leon. 2011. "Taiwan Faces Tough Water Choices." *The Guardian,* June 24.

Kemp, Rene, Saeed Parto, and Robert B. Gibson. 2005. "Governance for Sustainable Development: Moving from Theory to Practice." *International Journal of Sustainable Development,* 8(1/2): 12–30.

Keupp, Marcus Matthias, Angela Beckenbauer and Oliver Gassmann. 2009. "How Managers Protect Intellectual Property Rights in China Using De Facto Strategies." *R&D Management* 39 (2): 211–224.

Kim, Min-Jeong and Jai S. Mah. 2009. "China's R & D Policies and Technology-intensive Industries." *Journal of Contemporary Asia* 39(2): 262–278.

DOI: 10.1057/9781137358509

Klok, Chris and Tiehan, Zhang. 2008. *Biodiversity and its Conservation in China*. Alterra-rapport 1733.

Kreng, Victor B. and Chi-Tien Yang. 2011. "The Equality of Resource Allocation in Health Care under the National Health Insurance System in Taiwan." *Health Policy* 100: 203–210.

Krugman, Paul. 1991. *Geography and Trade*. Cambridge, MA: MIT Press.

Krugman, Paul. 1999. "The Role of Geography in Development." *International Regional Science Review* 22: 142

Kuo, Grace. 2011. "Understanding and Protecting Taiwan's Biodiversity." *Taiwan Today*, July 1. http://www.taiwantoday.tw /ct.asp?xItem=142518&CtNode=427

Lai, Dejian. 2003. "Principal Component Analysis on Human Development Indicators of China." *Social Indicators Research* 61(3): 319–330.

Lai, Hongyi. 2012. "Taiwan-Mainland China Energy Ties: Cooperation and Potential Conflict." In Wakefield *Taiwan's Energy Conundrum*. Woodrow Wilson International Center for Scholars Special Report No. 146.

Lan, Lawrence W. 2005. "Sustainable Development and Transportation: A Taiwan Perspective." Lecture note prepared for the Sustainable Metabolic Systems of Water and Waste for Area-based Society of the Socio-environmental Engineering Group, Hokkaido University, Japan, February 4.

Lau, Maren. 2003. "Coastal Zone Management in the People's Republic of China: A Unique Approach?" *China Environment Series* 6: 120–124.

Lee, Shsh-Der. 2010. "Fiscal Policy and Tax Reform in Taiwan." January 26, http://www.ttc.gov.tw/public/Data/025142321542.pdf.

Lee, Yu-Feng. 2008. "Economic Growth and Income Inequality: The Modern Taiwan Experience." *Journal of Contemporary China* 17(55): 361–374.

Leu, Horng-Guang. 2008. *Water Pollution Control Policies, Remediation Problems and Challenges*. Water Industry Development and Promotion Association. http://www.water.org.tw/simply/twic /PDF/A1–2Water%20Pollution%20Control%20Policies,%20 Remediation%20Problems%20and%20Challenges.pdf

Li, He. 2000. "Political Economy of Income Distribution: A Comparative Study of Taiwan and Mexico." *Policy Studies Journal* 28(2): 275–291.

DOI: 10.1057/9781137358509

Li, Bingqin and David Piachaud. 2006. "Urbanization and Social Policy in China." *Asia-Pacific Development Journal* 13(1): 1–26.

Li, Xiaoying. 2011. "How Does China's New Labor Contract Law Affect Floating Workers?" Harvard Law School Working Paper.

Li, Li and Wei-zhong Zeng. 2010. "Research Progress of Land Ecological Security Evaluation in China." *Journal of Geography and Geology* 2(1): 48–59.

Liang, Te-Hsin and Yu-Li Liu. 2006. *Internet Broadband Usage Adoption in Taiwan: An Overview Report.* Report for APIRA Conference, January.

Lim, Burton K., Judith L. Eger, A. Townsend Peterson, Mark B. Robbins, Dale H. Clayton, Sarah E. Bush, and Rafe M. Brown. 2008. "Biodiversity in China: Lost in the Masses?" *Harvard Asia Quarterly* 11: 12–23.

Lin, Chen, Ping Lin and Frank Song. 2010. "Property Rights Protection and Corporate R&D: Evidence from China." *Journal of Development Economics* 93: 49–62.

Lin, Chuan-Yao. Shaw C. Liu, Charles C.-K. Chou, Saint-Jer Huang, Chung-Ming Liu, Ching-Huei Kuo, Chea-Yuan Young. 2005. "Long-Range Transport of Aerosols and Their Impact on the Air Quality of Taiwan." *Atmospheric Environment* 39: 6066–6076.

Lin, Chun-Hung. 2003. "International Influences and the Transformation of the Telecommunication Regulations in Taiwan." *Chinese Journal of International Law.* 2(1): 267–288.

Lin, Chun-Hung. 2010. "Selected International Rules of Foreign Direct Investment in the Telecommunications Sector and its Influences on Taiwan's Telecommunications Legislation." *Annual Survey of International and Comparative Law* 16(1); 27–62.

Lin, Chun-Hung A. and Peter F. Orazem. 2003. "Wage Inequality and Returns to Skill in Taiwan, 1978–96." *Journal of Development Studies* 39(5): 89–108.

Lin, Tin-Chun. 2004. The Role of Higher Education in Economic Development: An Empirical Study of Taiwan Case. Journal of Asian Economics 15(2): 355–371.

Lin, Wen-Shyong, Chien-Hung Chen, Pei-Lin Chang, Kon-Tsu Kin, Shiuh-Tyng Tseng, Kuo-Hung Lee, Chun-Wei Chen. 2007. *Review of Water Use and Water Conservation Technology in High-Tech Industry.* Water Industry Development and Promotion Association. http://www.water.org.tw/simply/twic/PDF /C3–2Review%20of%20Water%20Use%20and%20Water%20

DOI: 10.1057/9781137358509

Conservation%20Technology%20in%20High-Tech%20Industry. pdf

Lipton, Michael and Qi Zhang. 2007. "Reducing Inequality and Poverty During Liberalisation In China: Rural And Agricultural Experiences And Policy Options." PRUS Working Paper 37.

Liu, Jin and Bin Lin. 2012. "Government Auditing and Corruption Control: Evidence from China's Provincial Panel Data." *China Journal of Accounting Research* 5: 163–186.

Liu, Lee. 2008. "Sustainability Efforts in China: Reflections on the Environmental Kuznets Curve through a Locational Evaluation of 'Eco-Communities.'" *Annals of the Association of American Geographers* 98(3): 604–629.

Liu, Ts'ui-jung. 2011. "Climate Changes and Water Resources in Taiwan History." PNC Conference Bangkok, October 19–21.

Liu, Jianguo, Zhiyun Ouyang, Stuart L. Pimm, Peter H. Raven, Xiaoke Wang, Hong Miao, Nianyong Han. 2003. "Protecting China's Biodiversity." *Science* 300: 1240–1241.

Liu, Xuejun, Lei Duan, Jiangming Mo, Enzai Du, Jianlin Shen, Xiankai Lu, Ying Zhang, Xiaobing Zhou, Chune He, Fusuo Zhang. 2011. "Nitrogen Deposition and its Ecological Impact in China: An Overview." *Environmental Pollution* 159: 2251–2264.

Liu, Yansui, Fugang Zhang, and Yingwen Zhang. 2009. "Appraisal of Typical Rural Development Models during Rapid Urbanization in the Eastern Coastal Region of China." *Journal of Geographical Science* 19: 557–567.

Lo, Szu-chia Scarlett. 2012. "Innovation and Patenting Activities at Universities in Taiwan: After Bayh-Dole-Like Acts." *World Patent Information* 34: 48–53.

Lo, Yueh-Hsin, Yi-Ching Lin, Juan A. Blanco, Chih-Wei Yu and Biing T. Guan. 2012. "Moving from Ecological Conservation to Restoration: An Example from Central Taiwan, Asia." Chapter in *Forest Ecosystems – More than Just Trees*, eds. Juan A. Blanco and Yueh-Hsin Lo. Shanghai: InTech Publishing.

Lyons, David. 2005. "Environmental Protection in Taiwan: Is it Too Much Too Fast?" *Journal of Contemporary Asia* 35(2): 183–194.

Ma, Jian. 2013. "China's Brutal One-Child Policy." New York Times Opinion Pages, May 21.

Mabogunje, Akin L. 1995. "The Environmental Challenges of Sub-Saharan Africa," *Environment* 37(4): 31–35.

DOI: 10.1057/9781137358509

McGregor, Peter G. and Swales, J. Kim and Turner, Karen. 2004. "The Impact of the Scottish Economy on the Local Environment: An Alternative to the Ecological Footprint?" *Quarterly Economic Commentary*, 29 (1): 29–34.

Ministry of Economic Affairs. 2012. "What's New." March 28. http://www.moea.gov.tw/Mns/english/news/News.aspx?kind=6&menu_id=176&news_id=25024.

Ministry of Education. 2011. "An Educational Overview." http://english.moe.gov.tw/ct.asp?xItem=12487&CtNode=2003&mp=1

Ministry of Foreign Affairs. 2010. "International Organizations." April 23. http://www.taiwan.gov.tw/ct.asp?xItem=27190&ctNode=1922&mp=1001

Mitchell, Richard and Frank Popham. 2008. "Effect of Exposure to Natural Environment on Health Inequalities: An Observational Population Study." *Lancet* 372: 1655–1660.

Mo, Yan-Chih. 2011. "Taipei, New Taipei City Launch Tamsui River Clean-Up Body." *Taipei Times*, August 11.

Mol, Arthur P.J. 2011. "China's Ascent and Africa's Environment." *Global Environmental Change* 21: 785–794.

Mon, Wei-The. 2001. "The Empirical Research on Crime Control Policy – -Taiwan Experience." Paper Prepared for the 2001 Annual Symposium on "Crime and Its Control in Greater China," Center for Criminology, The University of Hong Kong July 5~6.

National Bureau of Statistics. 2012. *China Statistical Yearbook 2011.* Beijing: China Statistics Press.

National Council for Sustainable Development. 2010. *2010 Annual Report on National Sustainable Development.* http://sta.epa.gov.tw/nsdn/en/PRINT/99Annual.pdf

National Statistics, Republic of China (Taiwan). 2013. "Yearly Statistics." Accessed May 20, 2013. http://eng.stat.gov.tw/.

Nooteboom, Sibout. 2007. "Impact Assessment Procedures for Sustainable Development: A Complexity Theory Perspective." *Environmental Impact Assessment Review* 27: 645–665.

Pan, Ke and Wen-Xiong Wang. 2012. "Trace Metal Contamination in Estuarine and Coastal Environments in China." *Science of the Total Environment* 421–422: 3–16.

Parks, Bradley C. and J. Timmons Roberts. 2008. "Inequality and the Global Climate Regime: Breaking the North-South Impasse." *Cambridge Review of International Affairs* 21(4): 621–648.

DOI: 10.1057/9781137358509

Paull, John. 2008. "The Greening of China's Food – Green Food, Organic Food, and Eco-labelling." Sustainable Consumption and Alternative Agri-Food Systems Conference Liege University, Arlon, Belgium, 27–30 May.

Peng, Ito and Joseph Wong. 2008. "Institutions and Institutional Purpose: Continuity and Change." *East Asian Social Policy. Politics and Society* 36(1): 61–88.

Piketty, Thomas and Nancy Qian. 2009. "Income Inequality and Progressive Income Taxation in China and India, 1986–2015." *American Economic Journal: Applied Economics,* 1(2): 53–63.

Qiu, Wanfei, Bin Wang, Peter J.S. Jones, Jan C. Axmacher. 2009. "Challenges in Developing China's Marine Protected Area System." *Marine Policy* 33(4): 599–605.

Rauch, Jason N. and Ying F. Chi. 2010. "The Plight of Green GDP in China." *Consilience: The Journal of Sustainable Development* 3(1): 102–116.

Roam, Gwo-Dong. 2005. "Taiwan's Protection of the Global Environment." *Harvard-Asia Pacific Review* 8(1): 14–16.

Science & Technology Policy Research and Information Center. 2010. *Taiwan Research Report.* Taipei: Science & Technology Policy Research and Information Center.

Shao, Kwang-Tsao, Ching-I Peng, Eric Yen, Kun-Chi Lai, Ming-Chih Wang, Jack Lin, Han Lee, Yang Alan, and Shin-Yu Chen. 2007. "Integration of Biodiversity Databases in Taiwan and Linkage to Global Databases." *Data Science Journal* 6: S2-S10.

Shay and Partners. 2011. *Telecommunications- Taiwan.* International Law Office Report, August 17.

Shealy, Malcolm and James P. Dorian. 2010. "Growing Chinese Coal Use: Dramatic Resource and Environmental Implications." *Energy Policy* 38(5): 2116–2122.

Shen, Bo, Lynn Price, Jian Wang, Michelle Li, and Lei Zeng. 2012. *China's Approaches to Financing Sustainable Development: Policies, Practices, and Issues.* Hoboken, NJ: Wiley Interdisciplinary Reviews: Energy & Environment.

Shi, Peijun, Jing Liu,2, Qinghai Yao, Di Tang, Xi Yang. 2007. "Integrated Disaster Risk Management of China." OECD Conference Hyderabad, Session III: Financial Management: Role of Insurance Industry, Financial Markets, and Governments, February 26–27.

DOI: 10.1057/9781137358509

Shi, Yao, Chuanbin Zhou, Rusong Wang, and Wanying Xu. 2012. "Measuring China's Regional Ecological Development through 'EcoDP.'" *Ecological Indicators* 15: 253–262.

Sicular, Terry, Yue Ximing, Li Shi and Björn Gustafsson. 2005. "The Urban-Rural Gap and Income Inequality in China." Paper prepared for UNU-WIDER Project Meeting "Inequality and Poverty in China," 26–27 August, Helsinki, Finland.

Smyth, Russell, Vinod Mishra, and Xiaolei Qian. 2008. "The Environment and Well-Being in Urban China." *Ecological Economics* 68: 547–555.

Stein, Karen. 2009. "Understanding Consumption and Environmental Change in China: A Cross-national Comparison of Consumer Patterns." *Human Ecology Review* 16(1): 41–49.

Sun, Julie C. L. 2012. "Taiwan Agricultural Technology Foresight 2025." Presentation posted on website www.biotaiwan.org.tw, July.

Taiwan Water Corporation. 2012. "About Us." Accessed July 28. http://www3.water.gov.tw/eng/

Tang, Zhiyao, Zhiheng Wang, Chengyang Zheng, and Jingyun Fang. 2006. "Biodiversity in China's Mountains." *Frontiers of Ecology and Environment* 4(7): 347–352.

Tie, Xuexi and Junji Cao. 2009. "Aerosol Pollution in China: Present and Future Impact on Environment." *Particuology* 7: 426–431.

Tsai, An-Yuan and Wen-Cheng Huang. 2011. "Impact of Climate Change on Water Resources in Taiwan." *Terrestrial Atmospheric and Oceanic Sciences* 22(5): 507–519.

Tsai, Kang-Ting, Min-Der Lin and Yen-Hua Chin. 2008. "Noise Mapping in Urban Environments: A Taiwan Study." *Applied Acoustics* 70: 964–972.

Tsai, Pei Chen and Shang-Hui Lin. 2012. "Taiwan's Path to Innovative R&D and Applications: Technology-Driven and Location-Driven Pilot Programs." *auSMT* 2(1): 7–10.

Tse, Pui-Kwan. 2010. "The Mineral Industry of Taiwan." In *2010 Minerals Yearbook Taiwan*. Washington, DC: US Department of the Interior.

Tse, Pui-Kwan. 2010. "The Mineral Industry of China." In *2010 Minerals Yearbook China*. Washington, DC. US Department of the Interior.

Tseng, Mao, Wei-Ta Feng, Chin-Tzu Chen, K. Douglas Loh. 2009. "Case Study of Environmental Performance Assessment & for Regional

DOI: 10.1057/9781137358509

Resource Management in Taiwan". *Journal of Urban Planning and Development*, September: 125–131.

Turner, Jennifer L. 2007. "In Deep Water: Ecological Destruction of China's Water Resources." In *Water and Energy Futures in Urbanized Asia: Sustaining the Tiger*, eds. Erik R. Peterson and Rachel Posner. A Report of the Global Strategy Institute at the Center for Strategic and International Studies.

UNDP. 2009. "Gini Index." *Human Development Report 2009*. http://hdrstats.undp.org/en/indicators/161.html

United Press. 2011. "Taiwanese Seas Threatened By Overfishing." Dec. 29. http://www.upi.com/Business_News/Energy-Resources/2011/12/29/Taiwanese-seas-threatened-by-overfishing/UPI-79181325170080/.

Vennemo, Haakon, Kristin Aunan, Henrik Lindhjem, and Hans Martin Seip. 2009. "Environmental Pollution in China: Status and Trends." *Review of Environmental Economics and Policy* 3(2): 209–230.

Wakefield, Bryce. 2012. *Taiwan's Energy Conundrum*. Woodrow Wilson International Center for Scholars Special Report No. 146.

Wang, Changcheng. 2012. "The Influence of Labor Market Development to Labor Relations in 21st and Measure of Labor Relations in China." Paper written for ILERA 16th World Congress, July.

Wang, Ching-Yu and Jhen-Bin Wang. 2010. "Analysis and Evaluation of Taiwan Water Shortage Factors and Solution Strategies." *Asian Social Science* 6(10): 44–67.

Wang, Jinnan, Hongqiang Jiang and Fang Yu. 2004. "Green GDP Accounting in China: Review and Outlook." UN Working Paper, at http://unstats.un.org/unsd/envaccounting/londongroup/meeting9/china_country_report_2004.pdf.

Wang, Jinxia, Jikun Huang and Scott Rozelle. 2010. "Climate Change and China's Agricultural Sector: An Overview of Impacts, Adaptation and Mitigation." International Centre for Trade and Sustainable Development Policy Brief No. 5, May.

Wang, Mark, Michael Webber, Brian Finlayson, Jon Barnett. 2008. "Rural Industries and Water Pollution in China." *Journal of Environmental Management* 86: 648–659.

Wang, Shuilin. 2008. "Overview." Chapter in *Public Finance in China: Reform and Growth for a Harmonious Society*, eds Jiwei Lou and Shuilin Wang. Washington, DC: World Bank.

WCED. 1987. *Our Common Future*. Oxford: Oxford University Press.

DOI: 10.1057/9781137358509

Weber, Christopher L., Glen P. Peters, Dabo Guan and Klaus Hubacek. 2008. "The Contribution of Chinese Exports to Climate Change," *Energy Policy* 36 (9): 3572–3577.

Wei, Binggan and Linsheng Yang. 2009. "A Review of Heavy Metal Contaminations in Urban Soils, Urban Road Dusts and Agricultural Soils from China." *Microchemical Journal* 94: 99–107.

Wei, Yiming, Hsien-Tang Tsai, Ying Fan and Rong Zeng. 2004. "Beijing's Coordinated Development of Population, Resources, Environment and Economy." *International Journal of Sustainable Development and World Ecology* 11: 235–246.

World Bank. 2012. "World Development Indicators Database." Databank.worldbank.org.

Wu, Tai-Yin, Azeem Majeed, and Ken N. Kuo. 2010. "An Overview of the Healthcare System in Taiwan." *London Journal of Primary Care* 3:115–19.

Xin, Tao and Chunhua Kang. 2012. "Qualitative Advances of China's Basic Education Since Reform and Opening Up." *Chinese Education and Society*, 45(1): 42–50.

Yang, Kuang-Ling. 2002. "Spatial and Seasonal Variation of PM10 Mass Concentrations in Taiwan." *Atmospheric Environment* 36: 3403–3411.

Yang, Lei. 2012." Improvement of Urban Water Environment of Kaohsiung City, Taiwan, by Ecotechnology." *Water Science and Technology* 66(4): 728–734.

Yang, Lin. 2007. "China's Three Gorges Dam Under Fire." *Time*, October 12.

Yao, Runming, Baizhan Li, and Koen Steemers. 2005. "Energy Policy and Standard for Built Environment in China." *Renewable Energy* 30: 1973–1988.

Yeh, Jiunn-rong. 2001, "Sustainable Development Indicators for Taiwan," Workshop on Sustainable Development Indicators, START, Chung-Li, Taiwan, Nov. 17–19.

Yeh, Kuo-liang. 2009. "The Diversity of Taiwanese Culture and Customs." Department of Chinese Literature, National Taiwan University Working Paper, http://www.uni-heidelberg.de/md/zo/sino/research/09_abstract-2.pdf.

Ying, Zhang, Minxue Gao, Junchang Liu, Yali Wen, Weimin Song. 2011. "Green Accounting for Forest and Green Policies in China — A Pilot National Assessment." *Forest Policy and Economics* 13(7): 513–519.

DOI: 10.1057/9781137358509

Yu, Ning. 2007. "Activities of Green Procurement in Taiwan. Presentation at Interactive Meet on Green Supply Chain and Green Purchasing," International Green Purchasing Network in Thailand, September 15–23.

Zha, Daojing. 2006. "China's Energy Security: Domestic and International Issues." *Survival* 48(1): 179–190.

Zhang, Haiyan, Michinori Uwasu, Keishiro Hara and Helmut Yabar. 2011. "Sustainable Urban Development and Land Use Change—A Case Study of the Yangtze River Delta in China." *Sustainability* 3: 1074–1089.

Zhang, Kun-min and Zong-guo Wen. 2008. "Review and Challenges of Policies of Environmental Protection and Sustainable Development in China." *Journal of Environmental Management* 88: 1249–1261.

Zhang, Xiaobo and Ravi Kanbur. 2005. "Spatial Inequality in Education and Health in China." *China Economic Review* 16(1): 189–204.

Zhang, Xiaodan. 2011. "China Environmental Labeling Program and Green Public Procurement." http://www.abntonline.com.br/Rotulo /Dados/Images/file/GEN%20AGM%202011%20-%20Linking%20 Ecolabelling%20with%20Government%20Green%20Procurement- Asian%20Experience(China%20).pdf

Zhao, Litao and Sixin Sheng. 2008. "China's 'Great Leap' in Higher Education." EAI Background Brief No. 394.

Zhou, Nan, Mark D. Levine, and Lynn Price. 2010. "Overview of Current Energy Efficiency Policies in China." *Energy Policy* 38(8): 1–37.

Zhou, Wei and Joseph S. Szyliowicz. 2006. "The Development and Current Status of China's Transportation System." *World Transport Policy and Practice* 12(4): 10–16.

Zhou, Yuan and Richard S. J. Tol. 2005. "Water Use in China's Domestic, Industrial and Agricultural Sectors: An Empirical Analysis." University of Hamburg Working Paper FNU-67.

Zhu, Wei Xing, Li Lu, and Therese Hesketh. 2009. "China's Excess Males, Sex Selective Abortion, and One Child Policy: Analysis of Data from 2005 National Intercensus Survey." *British Medical Journal* 338.

DOI: 10.1057/9781137358509

Index

DOI: 10.1057/9781137358509

DOI: 10.1057/9781137358509

CPSIA information can be obtained at www.ICGtesting.com
Printed in the USA
LVOW08*1506181113

361784LV00023B/314/P